Kipping Robert

Rudimentary Treatise on Masting, Mastmaking and Rigging

of Ship Also Tables of Spars, Rigging, Blocks

Kipping Robert

Rudimentary Treatise on Masting, Mastmaking and Rigging
of Ship Also Tables of Spars, Rigging, Blocks

ISBN/EAN: 9783337384371

Printed in Europe, USA, Canada, Australia, Japan

Cover: Foto ©berggeist007 / pixelio.de

More available books at **www.hansebooks.com**

RUDIMENTARY TREATISE

ON

MASTING, MAST-MAKING,

AND

RIGGING OF SHIPS.

ALSO,

TABLES OF SPARS, RIGGING, BLOCKS; CHAIN, WIRE, AND
HEMP ROPES, &c. &c.

RELATIVE TO EVERY CLASS OF VESSELS.

TOGETHER WITH

AN APPENDIX OF DIMENSIONS OF MASTS AND YARDS OF THE
ROYAL NAVY OF GREAT BRITAIN AND IRELAND.

By ROBERT KIPPING, N.A.,

Author of the Elements of Sailmaking, &c.

ILLUSTRATED WITH NUMEROUS WOODCUTS.

SIXTH EDITION.

LONDON:
JOHN WEALE, 59, HIGH HOLBORN.

1861.

LONDON:
BRADBURY AND EVANS, PRINTERS, WHITEFRIARS.

PREFACE.

In offering to the British public this rudimentary work, or outline of the Practice of Masting, Mast-Making, and Rigging of Ships, the author has no apology to offer for its production. It was composed in the hours of relaxation from official duties, during the period of his employment in one of the largest private Naval and Commercial Dock Establishments in the north of England, where he had frequent occasion to direct his attention to the mode of making masts and spars of various forms and dimensions, and witnessing other extensive works of rigging and equipment of ships. In addition to this, he devoted his evening hours to the perusal of the latest eminent publications on these subjects, and procured information from every possible source, with a view of rendering this work useful for elementary and practical purposes; he has, it is presumed, collected some valuable materials, and cast the result into one mould. A book of larger proportions and of greater price might have been drawn up with far less intellectual labour; this, however, would not have tended to accomplish the purpose which the publisher had in view, in issuing this Rudimentary Series, namely, to accord with the often limited resources of students. By adopting a small type and a full page, and joining together an immense quantity of tabular matter, the author has been enabled to attempt more, within the same number of pages, than has previously been effected in other works on the same subject.

The result of this undertaking is humbly submitted, trusting that it will meet with indulgence for any faults that may be contained therein; as, with the needful application to his profession, the author has no leisure to cultivate a literary style.

Since this work was written, Mr. Weale requested the insertion in these pages of Dimensions of Masts and Yards of some of II.M. ships, which were not originally intended. The author not wishing to disturb the work after it was arranged for the press, and yet being anxious to co-operate with the publisher to meet the suggestion and wishes of numerous applicants, it was proposed between them to add an Appendix to this work, which it is hoped will give satisfaction to the public.

CONTENTS.

CHAPTER VI.

CHAPTER VII.

CHAPTER VIII.

CHAPTER IX.

CHAPTER XIV.

APPENDIX.

MASTING AND RIGGING OF SHIPS.

CHAPTER I.

—◆—

General description of Masts.—Equipping a Ship with Three Masts and a Bowsprit.—The Timbers used for Masts.—Selecting Firs.

MASTS are long pieces of timber, rounded a great part of their length and erected on the keel of a ship, upon which are attached the yards, the sails, and the rigging, in order to their receiving the wind necessary for navigation. The lower masts of the largest ships are composed of several pieces of the soundest part of trees united into one body. As these are generally the most substantial parts of various trees, a mast formed by this assemblage is justly esteemed much stronger than one consisting of any single trunk, whose internal solidity may be very uncertain.

The principal things to be considered in equipping a ship with masts are, *first*, the number; *second*, their situation in the vessel; and *third*, their height above the water. The masts being used to extend the sails by means of their yards, it is evident that if their number were multiplied beyond what is necessary, the yards must be extremely short, that they may not entangle each other in working the ship, and by consequence their sails will be very narrow, and receive a small portion of wind. If, on the contrary, there is not a sufficient number of masts in the vessel, the yards will be too large and heavy, and cannot be managed without difficulty. There is a mean between these extremes which experience and the general practice of the sea have determined, by which it appears that in large ships every advantage of sailing is retained by three masts and a bowsprit.

The exact height of the masts in proportion to the form and size of the ship remains yet a problem to be determined. The more the masts are elevated above the centre of gravity the greater will be the surface of the sail which they are enabled to present to the wind,—so far an additional height seems to be advantageous. But this advantage is diminished by the vibrating movement of the mast, which operates to make the vessel stoop to its effort; and this inclination is increased in proportion to the additional height of the masts —an inconvenience which it is necessary to guard against : thus what is gained upon one hand is lost upon the other. To reconcile these differences, it is certain that the height of the mast ought to be determined by the inclination of the vessel, and that the point of her greatest inclination should be the term of the height above the centre of gravity. With regard to the general practice of determining the height of the masts, the extreme *breadth* of the ship from out to out has been admitted by long use the best rule for determining the length of the masts, that they may have proper support by the spread of the rigging.

The timbers commonly used for masts are fir and pine, and are distinguished by mast-makers by the name of the place from which they are exported—as, the Norway and Riga firs, Canada red, yellow, and white pines, &c.

The lower masts are generally made of yellow and the topmasts of red pine. In the selection of trees for making masts, yards, &c., it is of very great importance, not only on account of waste and expense, but because the safety of a ship in tempestuous weather frequently depends on its quality.

Before commencing to make a mast, yard, &c., or any part thereof, the tree designed for the purpose should be examined whether it be sound and fit, by cutting a short piece off the butt or thickest end, to see the heart. Should it have pale-red tints and white spots intermixed, and is rotten or shaky at the heart, the timber must have more pieces cut off while there remains sufficient length. When approved of the butt, examine along the sides, by taking chips off with the adze at different places and clear away the sap, and minutely examine every knot, rind-gall,* &c. If it possesses the necessary qualities of soundness line and measure it to the diameter and length required.

---

* Rind-gall is the damage which a tree receives when young, so that the bark or rind grows in the inner substance of the tree.

CHAPTER II.

Masting of Ships.—Placing of the Masts in relation to Water-line.—
The Proportions for the Rake.—Stive of the Bowsprit.

THE masting of ships, or the placing of the masts, belongs
to the business of the builder or constructor of the ship;
and the form given to the vessel varies the disposition of the
masts, for it is evident that a full-bowed ship requires her
foremast to be placed further forward than a sharp one;
consequently, though a general rule may be given, still every
builder should consider the nature of the form of his vessel,
and vary the disposition of her masts accordingly. The
following tables of rules, taken from Mr. Fincham's work,
will assist the builder in the placing of masts.*

LUGGER RIG.

Example 1 . . 55·0 ft. 16·75 ft. ,, 2 . . 77·0 ft. 22·7 ft. Species of Masts.	Known quantities.	Proportions in terms of known quantities.	
		Ex. 1. Common.	Ex. 2. Lugger.
Fore-mast before the middle	Length on water-line ×	Before. ·4	Before. ·396
Main-mast from the middle .	Do. do. ×	Abaft. ·037	Abaft. ·04
Mizen-mast or driver abaft .	Do. do. ×	Abaft. ·444	Abaft. ·396
Main-mast to rake .	In 12 feet . . ×	·16 in.	·12 in.
Fore-mast to rake .	Do. . . ×	·10 in.	·6 in.
Mizen-mast to rake .	Do. . . ×	·20 in.	·24 in.
Bowsprit to stive. .	Do. . . ×	·6 in.	·6 in.

* The load-water line is to be considered the principal line of
bearance of the ship, from which the stations of the masts are to be
determined.

LATEEN RIG.

Example . Length. Breadth. 125·0 ft. 32·0 ft. Species of Masts.	Known Quantities.	Proportions.
Main-mast from the middle	Length on water-line . ×	·000
Fore-mast before the middle	Do. do. . . ×	Before. ·407
Mizen-mast abaft . .	Do. do. . . ×	Abaft. ·407
Main-mast to rake . .	In 12 feet . . . ×	·000
Fore-mast to rake . .	Do. . . . . ×	Forward. ·23 in.
Mizen-mast to rake . .	Do.	Aft. ·12 in.

YACHTS, CUTTER RIG.

Yachts { Length. Breadth. Ex. 1. 63·1 ft. 19·2 ft. Ex. 2. 57·25 ft. 18·8 ft. Species of Mast.	Known Quantities.	Proportions in terms of known quantities.	
		Cutter. Ex. 1.	Cutter. Ex. 2.
Mast before the middle on the water-line . . .	{ Length of water-line taken from the fore part of the stem to after part of post . }	·112	·14
Mast to rake from the water-line . . .	In 12 feet . .	12 in.	15 in.
Bowsprit to stive from the water-line . . .	In 12 feet . .	7¾ in.	10½ in.
Bowsprit to house from the fore part of the stem	Breadth . .	·62	·53

SCHOONERS, THREE MASTS, BRIG FORWARD, COMMON, AND BERMUDA RIG.

	Length.	Breadth.
Schooner, three masts .	78·7 ft.	21·6 ft.
Do. brig for. Ex. 1 :	110·6 ft.	25·6 ft.
Do. do. Ex. 2 :	102·5 ft.	25·8 ft.
Do. common :	90·0 ft.	24·0 ft.
Do. Bermuda, Ex. 1 :	93·0 ft.	24·7 ft.
Do. do. Ex. 2 :	94·7 ft.	24·0 ft.

	Known quantities.	Proportions in terms of known quantities.					
		Schooners.	Schooners. Brig forward.		Common.	Schooners. Bermuda.	
		Three Masts.	Ex. 1.	Ex. 2.		Ex. 1.	Ex. 2.
Main-mast from the middle	Length of water-line	Abaft ·033	Abaft ·11	·107	·046	·103	·084
Fore-mast before do.	Do.	·295	·3	·294	·338	·279	·31
Mizen-mast abaft do.	Do.	·366	—	—	—	—	—
Main-mast to rake	In 12 feet	27 in.	33 in.	28 in.	24 in.	24 in.	33 in.
Fore-mast to do.	Do.	24 in.	28 in.	18 in.	15 in.	16 in.	36 in.
Mizen-mast to do.	Do.	30 in.	—	—	—	—	—
Bowsprit to stive	Do.	22 in.	36 in.	33 in.	34 in.	24 in.	22 in.

BRIGS.

Species of Masts, &c.	Known quantities.	Proportions in terms of known quantities.		
		Brigs of War.		Yachts as Brigs.
		Ex. 1.	Ex. 2.	
Main-mast abaft the middle	$\{$ Length on the water-line . . $\}$ ×	·147	·138	·144
Fore-mast before the middle	Ditto ×	·331	·323	·323
Main-mast to rake . .	In 12 feet . ×	10 in.	9 in.	10 in.
Fore-mast to rake . .	In 12 feet . ×	3 in.	2 in.	2¼ in.
Bowsprit to stive . .	In 12 feet . ×	51 in.	48 in.	52 in.

CORVETTES.[*]

Species of Masts.	Known quantities.	Proportions in terms of known quantities.	
		Ex. 1.	Ex. 2.
Fore-mast before the middle	$\{$ Length of water-line $\}$ ×	·372	·399
Main-mast abaft ,,	Do. ×	·079	·06
Mizen-mast abaft ,,	Do. ×	·375	·356
Rake of fore-mast . . .	In 12 feet . .	2 in.	1¾ in.
,, main-mast . . .	In 12 feet . .	6 in.	10½ in.
,, mizen-mast . . .	In 12 feet . .	10 in.	10¾ in.
Stive of bowsprit . . .	In 12 feet . .	64 in.	50 in.

[*] Corvettes embrace the tonnage of large merchant ships.

FRIGATES.

Species of Masts.	Known quantities.	Proportions in terms of known quantities.			
		Ex. 1.	Ex. 2.	Ex. 3.	Ex. 4.
Fore-mast before the middle . . .	{ Length of } { water-line } ×	·37	·364	·374	·39
Main-mast abaft do. .	Do. ×	·062	·073	·059	·068
Mizen-mast abaft do.	Do. ×	·341	·35	·335	·404
Rake of fore-mast .	In 12 feet ×	2 in.	1¾ in.	1¼ in.	1 in.
„ main-mast .	In 12 feet ×	6 in.	5¼ in.	5 in.	5 in.
„ mizen-mast .	In 12 feet ×	10 in.	11 in.	9 in.	9 in.
Stive of bowsprit .	In 12 feet ×	63 in.	54 in.	60 in.	60 in.

The preceding tables of the positions of the masts under different rigs are given, that constructors may form a comparison with other ships in their calculation of the centre of effort of the sails, and to bring the point of sail and the fore and after moments * within the limits, and according to rules laid down, and alter them as the results may require.

When a sharp-bowed vessel or ship has her mast to rake, it frequently eases her in pitching, but never adds to her sailing, the wind having less power on her sails; it is, however, necessary that a ship's main and mizen-masts should rake more than the fore-mast; for, by separating them in this way, the wind acts with more power on all the sails, when close-hauled, which otherwise would not be effected, and be of little or no advantage to the ship.

Before the positions of the masts are fixed, it is necessary to make a plan of the sails, and find the centre of effort, and compare it with data that have been furnished from ships that were found to work well. An example of the method of finding the centres of gravity of sails, and determining the position of the centre of effort of the moving force or the sails of a ship, is given, page 74, Part I., "Rudimentary Naval Architecture."

* The relation which the fore and after moments should bear to each other is 1 : 72 to 1 : 77. See "Elements of Sailmaking," by R. Kipping.

CHAPTER III.

—✦—

THE size of trees required for the making the lower masts
of the largest ships in the navy, both as to diameter and
length, but especially the latter, is so great that it is impossible to procure timber of natural growth fit for the purpose.
They are therefore composed of several pieces united into
one body. As before observed, these being generally the most
substantial parts of various tiers, a mast formed by this
assemblage is justly esteemed much stronger than one consisting of any single trunk, whose internal solidity may be
questionable. And the mode of securing these built masts,
as they are termed, by means of several strong hoops of
iron driven on the outside of the mast, seems to fulfil the
old adage of "a bundle of sticks that could not be broken
when so united." It will not be attempted in this rudimentary volume to describe how the several parts of a made
mast are put together, as it would require a very large plate
to accompany the explanation, and, besides would be of
little use to the *mercantile marine.* The method of making
single-tree masts is as follows :—

ON SINGLE TREE MASTS.

The stick, when appropriated, butted, and its soundness
examined, is laid upon the blocks or thawts, *t t t,* pieces of
plank piled upon one another, upon which it is to be trimmed;
when not quite straight, its hollow side is made the afterside; it is, however, usual to line the bent side first, and a
straight middle line is struck along it ; then the heights of
the decks set up thereon from the butt, which is called the
housing, and determines the place for the partners, *p,* where
the mast has the greatest diameter. From the heel, set up

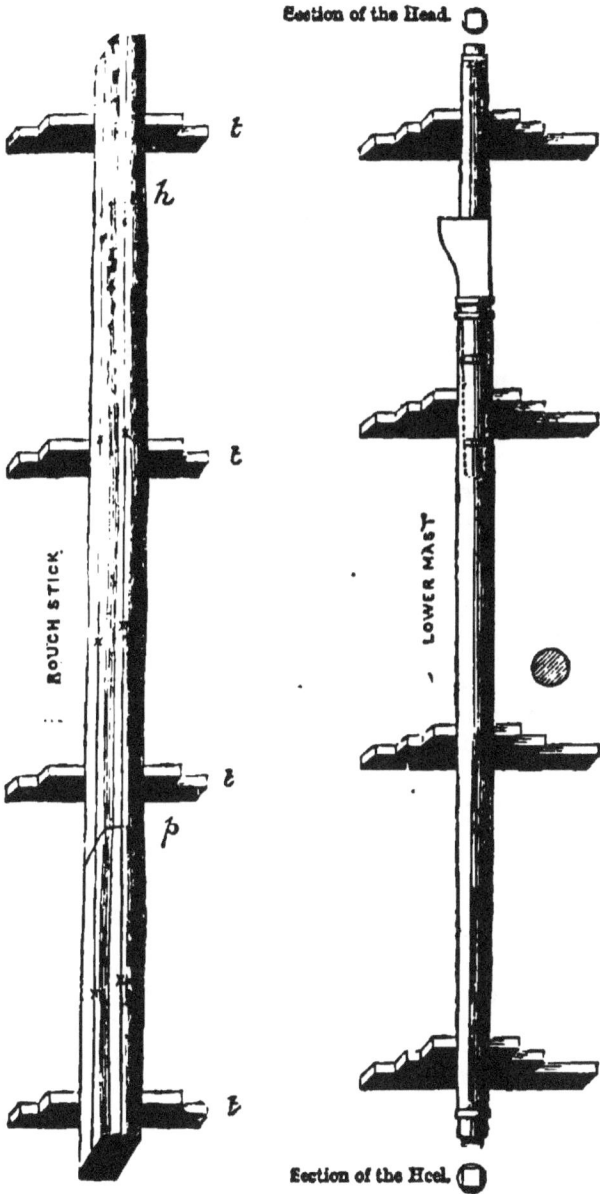

Section of the Head.

ROUGH STICK.

LOWER MAST.

Section of the Heel.

B 3

the whole length of the mast, and from the place for the partners the mast is divided into four parts each way, termed off-sets or quarters; that nearest the partners, the first; the next, the second; the other, the third. The different proportions of the given diameter are set off at those places: at the first, second, third quarters, heel, partners, hound, and head; the mast is formed to these the fore-and-aft way; but athwartship it is made straight from the third upper quarter to the head, instead of preserving the proportion given at the hounds, h. All masts are first made square by being hewed perpendicular to the lines of the upper surface; the sides are hewed in at different places until the plumb-line is perpendicular to the line struck on the surface. The rough wood between these places chopped is hewed off out of winding with the spots plumbed down.

When the mast is sided, it is canted or turned with the trimmed side up, and a straight middle line is struck along perpendicular to a vertical line upon each end; then the quarters and other divisions are squared up from the first middle line struck, and the diameters set off as before from the second middle line. The sides are then hewed square to the surface from those lines as done the first.

The parts that are to be rounded are eight-squared, thus—$\frac{2}{24}$ of the diameters are set off on each side of the middle line, on every side, or $\frac{1}{24}$ set in within the edges, and lines struck with fair curves: then the angles or edges are taken off straight to the lines on each side, and made cylindrical. At the stops $\frac{1}{8}$ of the given length, or two-thirds of the mast-head are left square for the hound-pieces, and above once and a half the depth of the trestle-trees for the trestle-trees, &c.; but when the mast has long hound-pieces, it is left square three-fourths of the length of the mast-head, above the lower part of the lower hoop has its angles rounded off in an easy manner; for by making it cylindrical, a proper seating for the cap would not be obtained. The angles are first taken off to one-seventh of the size of the mast-head each way; and then the other angles formed are taken off to one-fourth of the size of the squares, till they are reduced sufficiently small to form a fair curve with a plane; and below the square it is formed into the round, or what is called *hanced into it*, with a hance about five-sixths of the length of the hounds, below which it is made round—but sometimes a small chamfer only is taken off the masthead.

THE HOUND-PIECES.

The annexed diagram exhibits the fittings of the main-mast to brigantines, and the hound-pieces and knees in one piece:—

a a, cross-trees.

b, short cross-tree.

c, the lower-cap with an iron band.

d, bolster.

e, iron plate with an eye bolt.

f, for boom topping lift.

g, eyebolt for the forebrace.

h, hoop and large eye for throat hallyards (this bolt is commonly fastened with a nut and screw at the fore part).

i i i, masthead battens.

k k k, eyebolts through the hoops for peak-hallyards.

m, sheave in masthead for top rope.

The hound-pieces in single-tree masts are formed with the knees in one piece,—they are in thickness half that of the trestle-trees, and never less than three inches thick ; in breadth they are the same as the masthead, with an extra breadth for the breadth of the knee which is formed on the fore part for supporting the trestle-trees, the breadth of the knee being equal to the diameter of the topmast, or it may extend to the fore side of the fid-hole ; and their length is two-fifths of the length of the head, without the additional length, or one-eighteenth the given length of the mast.

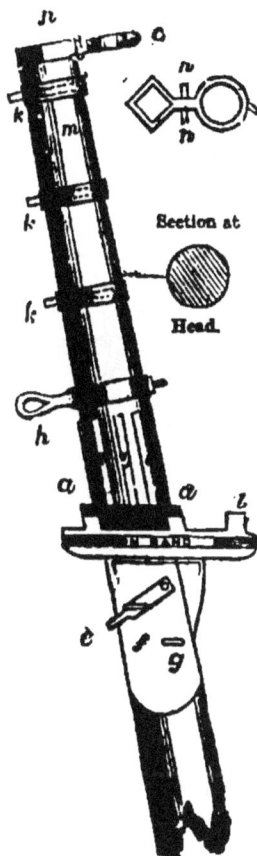

MAST-HEAD WITH OAK CHEEKS.

To this mast-head is shown the fittings, as is common to brigantines' foremast. In the adjoining figure—

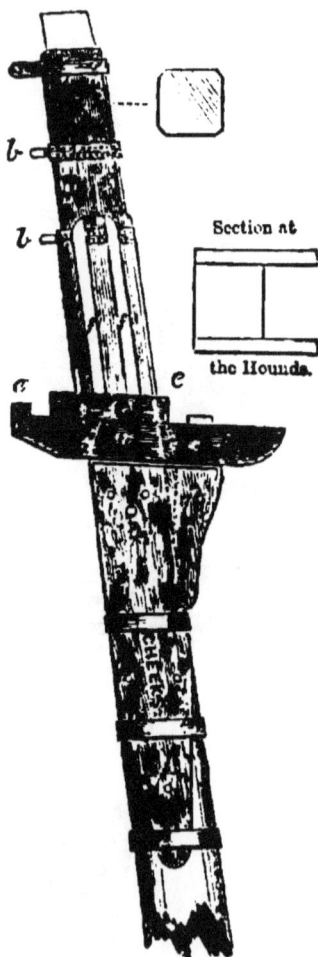

a, the hoop and roller for horizontal stay.

b b, eyebolts through the hoops, clenched on the fore side for peak-hallyards.

c, bolster.

d, trestle-trees.

e e, cross-trees.

f f, battens on the mast-head.

The hound-pieces are coaked to the mast either by two square coaks formed out of the mast, or by circular coaks; and are bolted with five through bolts, driven through from alternate sides, and clenched upon the opposite. They are placed, the two upper and the two lower bolts, about one-fifth the diameter of the masthead from the fore and after sides; the two upper bolts about nine inches from the stops, and the two lower one-third the length of the hounds from the lower end; the other bolt is placed in the middle line of the mast, and at equal distances from the upper and lower bolts. A strengthening bolt is driven fore and aft through the knee and hounds-piece, just above the two upper bolts. The lower end is nailed with about six nails, two inches from, and following the curve of the end; a hoop is, however, to be preferred on the lower end of the hounds-piece, as the nails injure the mast.

THE FITTINGS OF THE MAST-HEAD AND PUTTOCK-SHROUDS.

The fittings which are commonly fixed for mast-heads, are shown in the figs. 1 and 2. In the royal dockyards, a chain necklace is preferred to a hoop, as it brings less strain on the mast; this chain has shackles, *a b*, fitted into the links to which the puttock-shrouds are attached, *s s s s*.

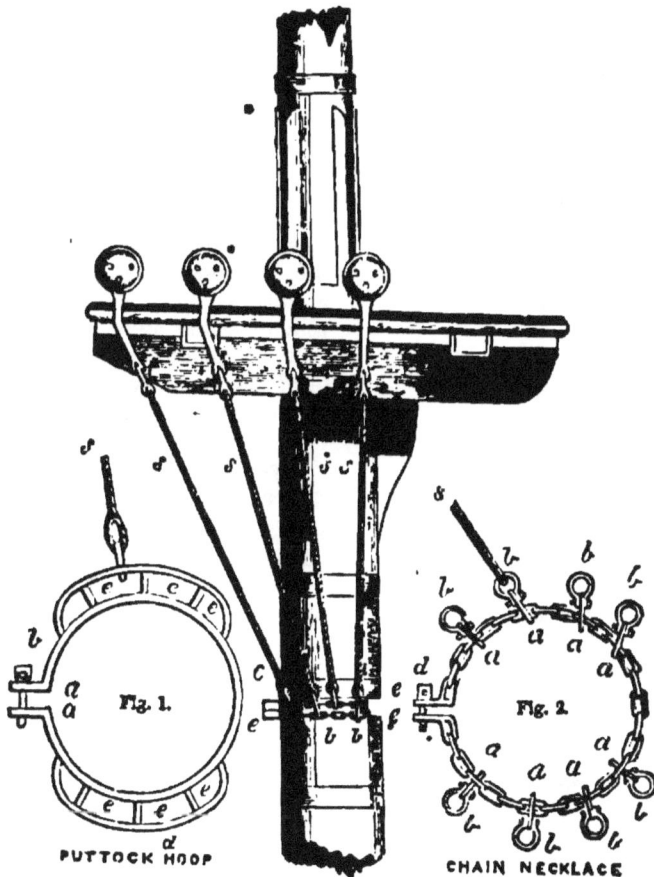

PUTTOCK HOOP   Fig. 1.   Fig. 2.   CHAIN NECKLACE

It is brought tight on the mast by a screw, *d*, passing through the ears, *e e*. But in the merchant service, a hoop is fitted, and the puttock-shrouds are attached, as *c d*, fig. 1, and kept separate by the bars, *e e e*, which also supports the hoop. It is similarly brought tight on the mast, as the chain necklace, by having two ears, *a a*, and the screw *b*.

THE FITTINGS FOR THE FORE-MAST-HEADS OF STEAM VESSELS.

In the annexed sketch the fittings which form the cross-trees are shown thus :—

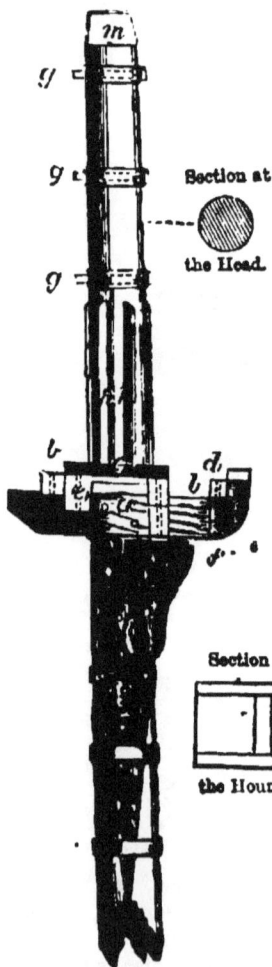

Section at the Head.

Section at the Hounds.

a, the trestle-trees.
b b, the cross-trees.
c, bolster.
d, connecting piece.
e, block fitted on the after side of the mast, in which are two sheaves for throat-hall-yards.
f, eyebolt for slinging the foreyard.
g g g, eyebolts for peak-hallyards.
h h, battens on the mast-head.

THE FITTINGS FOR THE MAIN-MAST-HEAD OF STEAM VESSELS.

In the annexed figure are shown the several fittings, as,—

a, trestle-trees.

b b, the cross-trees.

c, the bolster, which reaches from the fore-side of the mast to the chock, *g*, between the trestle-trees.

d, block with two sheaves, as shown in *e*, foremast, for throat-hallyards.

e, plate and eye for the boom-topping-lift with two bolts.

f, eye-bolt for the topsail brace.

h, eye-bolt for the fore-brace.

i, tenon for iron cap.

k, sheave in mast head for the top rope.

n-n n, eye-bolts for the peak-hallyards.

M, iron plate round the mast for the chain rigging.

The head of this mast is round, as per the section, and the hounds are made square; the knee and hounds-pieces are in one piece, and secured with three bolts, also a hoop is put on at the lower part with two bolts drove through it.

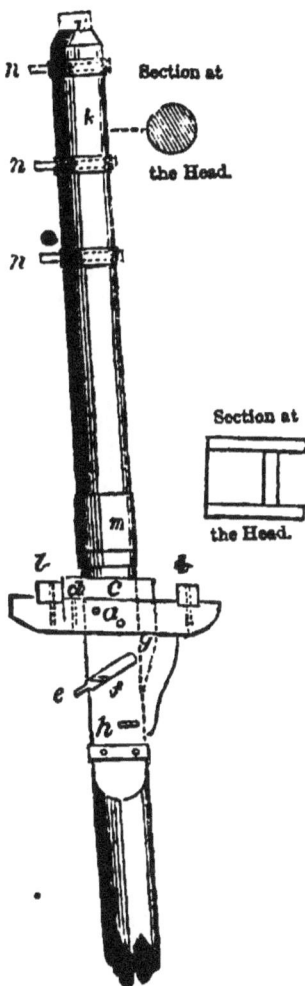

Section at the Head.

Section at the Head.

BRIGANTINE'S FORE-MAST AND MAIN-MAST.

The fore and main masts, as shown in pages 11 and 12,
have eye-bolts, *k k k*, for the peak-hallyard, and the main
mast has an eye and outrigger for the throat-hallyards, with
a plate for the boom-topping-lift, *e*. The outrigger is bolted
through the mast, and secured with nut and screw at the
point.

THE FRAMING OF THE MAST-HEADS OF STEAM-VESSELS.
FORE-MAST.

The fore-mast has two cross trees, *c c c*; their lengths
are, for the foremost one, *one-third* the hounded length of the
topmast, and the after one *two feet* longer than the foremost
one. They are curved aft, nearly equal to the diameter of

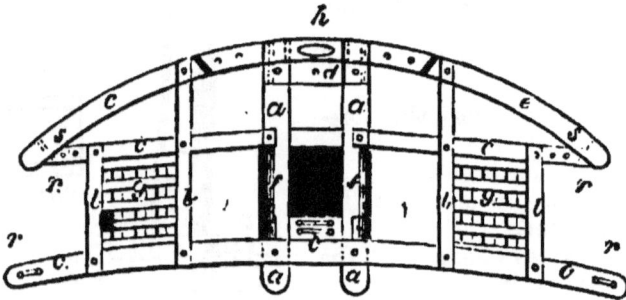

the topmast. The framing is formed by the trestle-trees,
a a, and cross-trees, and have, in connection with these, the
curved cross-tree, *e e*, which is joined on the fore end of the
trestle-trees, and forming a sweep to the outer end of the
foremost cross-tree, its after side having a connecting
piece, *d*, bolted on, and an iron strap, *s*, over each end. The
gratings, *g*, between the cross-trees are fixed and strength-
ened by two iron plates, *b b*, for standing upon. Rollers, *r r*,
are fixed in the ends of the cross-trees, for the topmast
rigging, and a block between the mast and after cross-tree,
for the throat-hallyards; between the fore cross-tree and
connecting piece a hole, *h*, is made for the slings of the yard,
and the bolster, *f*, is brought on the trestle-tree for the lower
rigging to lie over.

THE FRAMING OF THE MAST-HEAD OF MAIN-MAST OF STEAMER.

Two cross-trees, *a a*, and two trestle-trees, *b b*, combined together is the framing of the main-mast. The cross-trees are in length that the foremost one may carry the top-mast shroud *two feet* without a straight line from the stops to the outer part of the gunwale, and the after one is longer than the foremost one by two feet. Sufficient space is to be allowed between the cross-trees for the block, *c*, for the throat-hallyards, diameter of the topmast, *t*, and one inch for the heeling. The breadth of the trestle-trees extends the same before the foremost cross-tree as abaft the after one. Rollers, *r r*, are fixed in the ends of the cross-trees for the topmast rigging, and bolsters, *f f*, are brought on the trestle-trees for the lower rigging.

Small steamers have but one iron cross-tree.

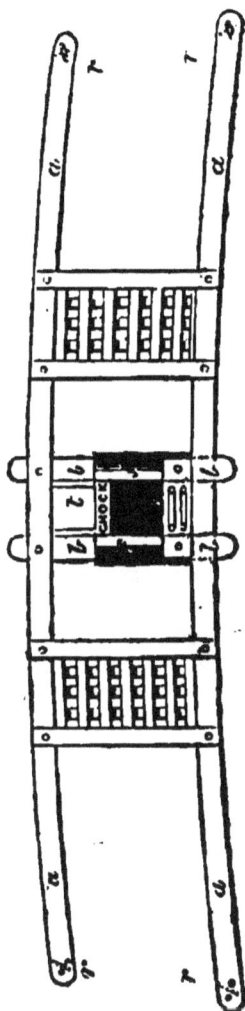

THE FRAMING OF THE MAST-HEAD OF BRIGANTINE'S MAIN-MAST.

The framing consists of two long cross-trees, c c, and a cross-tree, b; the long cross-trees are separate the thickness of the mast-head at the stops, and the short cross-tree is beyond the foremost cross-tree, the diameter of the heel of the topmast, t, and one inch for the heeling. The length of the

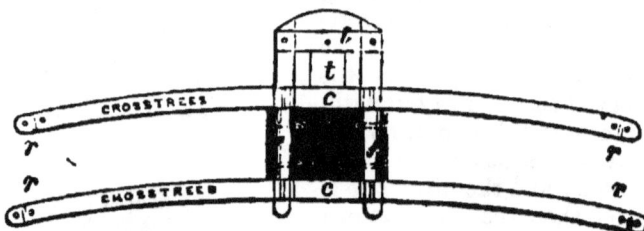

two long cross-trees, the same, must be observed as for the mainmasts of steamers. A chock is placed between the trestle-trees at the fore part of the short cross-tree, to receive an iron band, c, page 11, which passes round to the after part of the trestle-trees. Rollers, r r, are fixed in the ends of the cross-trees for the topmast rigging, and likewise there are bolsters, f f, for the lower rigging.

THE FITTINGS OF SCHOONERS' AND CUTTERS' MASTS.

The given lengths of schooners' and cutters' masts are generally the heading and hounding, and of similar form as single-tree masts without cheeks, but differing in having their heads rounded, and the diameter at the hounds or stops about a quarter less than the given diameter; a stop is formed about one inch on the foreside at the hounds, for the support of the lower cap, d.

On the masthead is placed five hoops; the lower hoop, a, is made with a wide collar to receive the shoulder of an iron outrigger, with an eye formed for the throat-hallyards; this hoop is put on from 2 feet to 2 feet 6 inches above the underside of the lower cap. The upper hoop is placed 6 inches below the upper cap, and three others are spaced at equal distances between the upper and lower hoops. These hoops, *h h h,* have all eyebolts drove through them from the afterside of the mast, and clenched on the foreside for the peak-hallyards; their eyes lie horizontally; they are placed on the head for the upper and lower ones to be in the middle line on the afterside of the mast, and the two

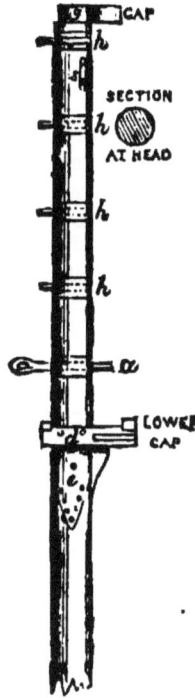

CAP
h
SECTION
h
AT HEAD
h
h
a
LOWER CAP

between these 1½ inch on each side of the middle line. A hoop is also driven on the heel of the mast, about 6 inches above the shoulder of the tenon. There is a sheeve, *s,* in the masthead for the top rope; *g,* is the iron cap; *d,* the lower cap; *e,* the hounds-piece. The afterside of the mast is coppered in the wear of the gaff and boom.

CHAPTER IV.

Bowsprits of Single Trees and Caps.—On the Jibboom.—Saddle for the Jibboom. —On the Flying Jibboom.—The jibboom and Flying Jibboom in one.—Top-masts.—Top-gallant and Royal-masts in one.—Stumppole top-gallant-masts.—Caps on the Lower Masts.—Top-mast Cap.

THE given length of a bowsprit is from the fore part of the tenon of the cap to the after part of the tenon at the heel. The part which rests upon the stem and apron is called the *bed;* the inner part, from the outer end of the bed to the heel, the *housing;* and nearly at the outer end is the head, or *bees' seating.* The general proportion of the outer end to the given diameter is *two-thirds,* and the inner end *five-sixths* of the given diameter. The ends are tapered from the bed each way, and rounded their whole length, excepting the bed, and on the upper part from the bed to the outer end is left square, for the men to walk out upon. Cleats used to be nailed on the bowsprit for the collars of the stays and bobstays; instead of these, wood is now left on to form the stops, *a a a,* as here shown.

There has been lately introduced another method of making these stops, which are, two iron straps or splints, *b b,* with stops, *s s s,* formed on them; they are let in flush on the under-side, and extend from the forestay-hoop to the outside of the cap, and therefore adds materially to the fastening of the cap, and preventing the under part of it from working loose. These straps are two inches wide and a quarter of an inch thick in the largest size single-tree bowsprit; the stops on them are about the thickness of the hoops for the bobstays, and the ends of them screw-bolted, to pass through the lower part of the cap to screw it up firmly to the end of the bowsprit. A few nails fasten the straps until the hoops are put on to their places.

To hoop a single-tree bowsprit.—One hoop is driven on the heel, and one 4 inches within the aftside of the cap.

The *bees* are fixed on the head, as shown by *g*, and have sheeve holes for the foretopmast-stay.

The *heel*.—The lower part of the heel is on a level with the deck, and the upper part in the direction of the bowsprit.

The *bowsprit cap.*—The length of the cap is in general five times the diameter of the jibboom, and the breadth twice the diameter, and the thickness one inch less than the diameter of the jibboom. The cap is of a parallel thickness, and the athwartship sides are cut square and parallel to the breadth. The ends are square athwart and fore-and-aft to the angle made by the stive of the bowsprit.

Below the upper end, in the direction of the stive, its thickness is set down on both sides, and a line, A A, drawn through parallel to the upper part of the cap, which is the station for the upper part of the hole for the jibboom. Set down square to the line A A, the diameter of the jibboom B, and three quarters of an inch play; draw the line c c, parallel to A A; this will give the station for the lower part of the hole. Draw A a and c c square across athwartship of the cap. As the hole will be an oval shape, both sides athwartships, on account of the stive, the distance A c will therefore be the long diameter, and the perpendicular B, equal to *b b*, the short diameter; to these diameters an ellipse or oval must be described for the hole that the jibboom slides through. Observe, in setting off the hole, that three quarters of an inch is to be added to the size for leathering. From the under part of this hole, set down two-sevenths the diameter of the bowsprit square across, as G *g*, for the upper part of the hole, and draw a line, G G, parallel to c c; then five-sixths the diameter of the outer end of the bowsprit is set down the middle line F F; and E *e* and E E are respectively drawn parallel to G *g* and G G; these lines will give the lower part of the square hole on the fore and after parts of the cap.

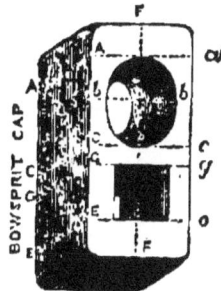

The cap, when trimmed and prepared for the tenon on the outer end of the bowsprit, is next iron-bound and eye-bolts driven through and clenched, for the man-ropes, foot-ropes, &c.

ON THE JIBBOOM.

The jibboom has a straight line struck across the middle : the butt end of the tree is worked inwards, and the length set up. The given diameter is at the bowsprit cap, or at one-third the length from the inner end, from which it is made parallel. From the cap to the outer end is divided into four quarters, and the outer end is made two-thirds the given diameter. It is lined to the size, and hewed plumb ; afterwards squared and eight squared, and three and a half diameters is left eight-squares from each end, and between the ends it is rounded. A stop is made at the outer end once and one-sixth the diameter in length ; and a sheave hole one diameter and one-sixth in length within the stop from the upper side, for the outhauler or the jibstay to pass through ; and at the heel is a horizontal hole for the lashing down of the heel ; a horizontal sheave once and one-sixth the diameter in length, for the top-rope. An iron for the flying-jibboom is placed upon the starboard eight-square at the outer end.

SADDLE FOR THE JIBBOOM.

The saddle is fayed upon the bowsprit, at one-third the length of the jibboom within the outer end. It is in length one-half the diameter of the bowsprit ; in width, one-half the diameter of the jibboom ; and, in thickness, one-sixth the given diameter. A seat is made upon the upper part for the heel, so that the jibboom may lie parallel with the middle line of the bowsprit, and it is fastened to the bowsprit after the bowsprit and the jibboom are rigged, with one rag-pointed bolt in the centre and a nail in each end.

ON THE FLYING JIBBOOM.

The heel of the flying-jibboom generally steps against the cap ; the usual proportion is two-thirds the jibboom and two-thirds the given length of the flying-jibboom. The given diameter is at the iron on the end of the jibboom, and the outer end is two-thirds, and the inner end three quarters of the given diameter. From the place of the given diameter it is made parallel to two-thirds the length from the inner end; the distances from these to the ends are rounded. A stop is made to one diameter in length within the outer end, and a vertical sheave is cut within the necking, for the flying-jibstay; and at the inner end a horizontal hole is bored.

THE JIBBOOM AND FLYING-JIBBOOM IN ONE.

Here the length to the stops will be the length of the jibboom; and from the stops to the outer end will be two-thirds the given length of the flying-jibboom. The diameter at the stop for the jibboom is two-thirds, and at the extreme end one-third of the given diameter. It has sheaves cut for the jib and flying-jibstays as the preceding jibbooms.

TOP-MASTS.

The length of topmasts is given the same as the standing masts, viz., the hounded and headed lengths; or the whole length including the length of the head. They require to be such thickness as to permit the hounds to pass through the cap. The given diameter is at the lower cap, from which place it is made parallel to the heel; and, at the stops and head, the diameters are four-fifths and five-ninths respectively of the given diameter. It is lined to these sizes, and *hewed* from the lower cap sixteen-square, and rounded to the under part of the hounds, which are nearly eight-squared, to admit them through the round hole in the lower cap. The fore and after stops, *c*, are formed as per the figure annexed; the head is left square above the stops, and the edges chamfered between the upper side of the cross-trees and the under side of the cap. The heeling is to be square, and the edges chamfered; and, if not sufficient to fill the hole in the trestle-trees, fillings are fayed and nailed thereto to supply such deficiency, allowing a quarter of an inch play. The *heeling* is in length two to two and a-half diameter, and brought into the size of the eight-square. In the middle of the heeling a square hole, *f*, is cut through athwartships for the *fid*; its lower part to be the depth of the trestle-trees above the upper part of the hoop, and one inch for an iron plate that is brought on the trestle-trees. *Iron fids* are mostly used, and are in length once and a-half the given diameter of their lower masts; in depth one-third the given diameter of their topmast, and in thickness two-thirds their depth.

Fids are made square to the given dimensions, and one end rounded, the other snaped from the under side, and a hole for a laniard in both ends.

The sheave-hole for the top-rope.—There is one sheave, *s*, at the heeling above the fid-hole, which is cut through transversely in the middle of the eight-square of the larboard side to the foremost eight-square on the starboard side. A groove is taken out, rather larger than the top rope, when passing through the square hole in the trestle-trees, and is in the direction of the sheave-hole.

The sheave-hole for the topsail-tye.—Most of the ships in the merchant-service have a sheave-hole cut through the middle of the hounds fore and aft; but, as it weakens the mast, it is preferable to have *blocks* in lieu of a sheave-hole.

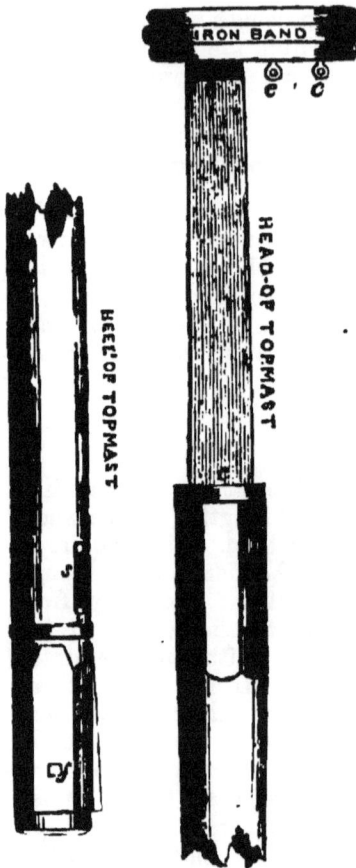

TOP-GALLANT AND ROYAL-MASTS IN ONE

It is now common to have the top-gallant and royal masts in one, T, r. The given length is set off from the lower end; and, from the heel to the length of the top-gallant mast is set off; then, from the top-gallant stops, the length of the royal mast is measured, and a very short pole above the royal stops, t, is allowed.

The given diameter, which is at the station of the cap, is set off from the lower end; and at the top-gallant stops, h, it is $\frac{10}{12}$ of the given diameter, and at the royal stops three-fifths of the given diameter. It is quartered and graduated the whole length, and rounded quite through, leaving only the lower part, a a, square.

The heeling.—The heeling is made so as to conform to the space between the cross-trees and trestle-trees, which is the same to the top-gallant mast, as has been explained for the topmast.

The fid-hole.—The lower part is one diameter and one inch up from the heel; for the up and down way half the diameter, and athwart-ships two-thirds of what it is up and down.

The sheave-hole for the top-rope.—There is one sheave-hole placed in the starboard foremost eight-square, with its lower part three diameters from the lower end, and in length one diameter and one sixth.

The sheave-holes for top-gallant and royal tyes.—Sheave-holes are cut in a fore-and-aft direction, half the diameter below each of the stops; the length of them is equal to the diameters at the stops and one sixth, and are lined with copper.

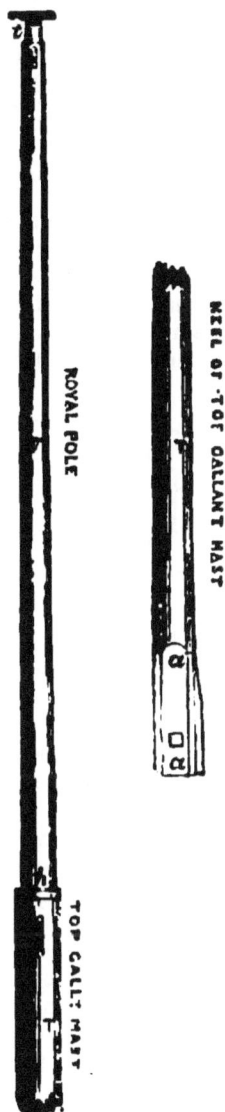

c

STUMP-POLE TOP-GALLANT MASTS.

Top-gallant masts are sometimes fitted with stump poles, when they frequently have a sliding gunter-mast fitted to them. The length is set off and formed the same as the top-gallant and royal masts in one, excepting that above the stop is a long pole head. The diameter of the extreme end of the pole is half the given diameter.

CAPS ON THE LOWER MASTS.

The principal caps of a ship are those of the lower masts, made of African oak, in the merchant service; they are in breadth equal to twice, and in thickness five-sixths the diameter of the top-mast. Two large holes are cut through them, the one square to fit on the lower mast-head, the other round, for the top-mast to slide through. The caps are trimmed or sawed to their dimensions, and their upper and under sides made straight and out of winding, and the ends are rounded off to an arc of a circle, the sides with a small curve. The holes are set off from the centre of the under side of the cap at equal distances; the substance of wood left between the holes to be half the taper of the mast head, and the thickness of the chock between the trestle-trees. The round hole is the foremost one for the top-mast, and is sweeped to seven-eighths of an inch larger than the diameter, to allow for the thickness of the leather, and one-eighth of an inch for play. The square hole is set off $\frac{9}{10}$ fore-and-aft, and $\frac{8}{10}$ athwart-ships, of the size of the tenon on the mast head, and tapered one inch to a foot, the fore part and each side three-eighths of an inch to a foot, and the after part square, towards the upper side of the cap: this is done for what is termed "strengthening down." The wood before the round hole is two-thirds the depth of the cap, and the wood left beyond the after part of the square hole is once the depth of the cap. The depth is usually reduced on the edge of the cap $\frac{1}{12}$, to make it as light as possible. The iron-hoop round the cap is commonly one-third its depth, and varies in thickness from one-fourth to five-eighths of an inch, according to the size of the cap.

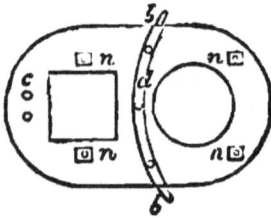

These caps have four eye-bolts, *n*, driven through the cap from the under side for the top-rope-pendants,&c. One of the bolts is placed on each side of the square hole near the edge of the cap on the under side, and one on each side of the round hole; at the fore part, with their eyes athwart-ships and well clenched upon an iron-plate let into the upper side of the cap. On the upper side is also a plate, *d*, for the lower lifts, with an eye in each end; it is in breadth about one-third the depth of the cap, and in thickness from one-quarter to five-eighths of an inch. Three bolts are driven to secure it on. All caps have horizontal strengthening bolts driven through them and clenched.

To let on the caps of the lower masts.—From the square hole cut in the cap, take the size at the lower side, and as the cap is not to go down to the shoulders on the first letting on within an inch and a half, for the shrinking of each, the size of the tenon is that distance set off; when this is done, take the size of the upper part of the hole and depth, and trim the head of the mast to the size. It is to be observed that all caps are to be raised above a level from the middle line on the mast to resist the weights that act on the fore side of them.

TOPMAST CAP.

The proportion of the topmast cap to the diameter of the

top-gallant mast is the same as the preceding to the topmast. Two holes are formed, the one square to fit the topmast head, the other round for the top gallant mast. They have four eye-bolts, *e e e e*, driven similar to the lower cap, and the cap is iron bound as described for the lower cap. See also sketch at page 24.

Several vessels in the merchant service have got *iron caps* fitted on the mast-heads and on the bowsprit-end. The cap of the bowsprit in this case is fitted on *square* on the end, and the hole is round for the jibboom going through. These *iron caps* look uncommonly snug and light, and seem to answer their purpose well.

CHAPTER V.

On Yards made of Single Trees.—On Main and Fore-yards.—On Topsail
Yards.—On Cross-jack Yards, Spritsail Yards, Top-gallant Yards, Royal
Yards, and Studding-sail Yards.—On the Driver, or Spanker, and Main
Booms.—On the Mizen and Main Gaffs.—On Trysail Masts.—Studding-
sail Booms.—Trestle-trees and Cross-trees for Lower Masts.—The Framing
of the Trestle and Cross-trees.—Tops of Merchant Ships.—Top-mast
Trestle and Cross-trees.

ON YARDS MADE OF SINGLE TREES.

Yards are either square, lateen, or lug-sail; the first are
suspended across the mast at right angles, and the two latter
obliquely. The square yards are of a cylindrical form,
tapering from the middle, which is termed the slings,
towards the extremities, which are called the yard-arms, and
at the slings is the place of the given diameter; the distance
between the slings and the yard-arms on each side is
quartered, which are distinguished into the first, second, and
third quarters, and yard-arms.

ON MAIN AND FORE-YARDS.

The middle quarters are formed into eight squares, the
after square to half the length of the yard, or to one foot
six inches beyond the quarter iron for after batten, and the
others, which are for the sling hoop and truss, one-eighth the
length of the yard, and from the eight squares in the middle
each side is rounded to the outer ends, except at the place
of the sheave-holes in their arms for the topsail sheets,
which are left square the length of the sheave-hole. It is
to be observed, however, that sheave-holes weaken the yards,
and an iron cheek block, *c*, is far superior, being brought
on the after side, even with the stops, *d*, for the topsail
sheet. .

The stops, *d d*, are formed out of the yards, in squaring
for the cheek blocks. Yards are fitted at their outer ends for
rigging out studdingsail-booms, with four *boom-irons ;* there

arc two on cach side of the yard; the outer one is named the *yard-arm-iron*, *b h*, and the inner one, which is placed at $\frac{1}{16}$ the length of the yard from the outer end, the *quarter-iron*, *n*. The outer boom-iron is composed of two parts, the strap, *g*, which is let in the thickness lengthways upon the yard-

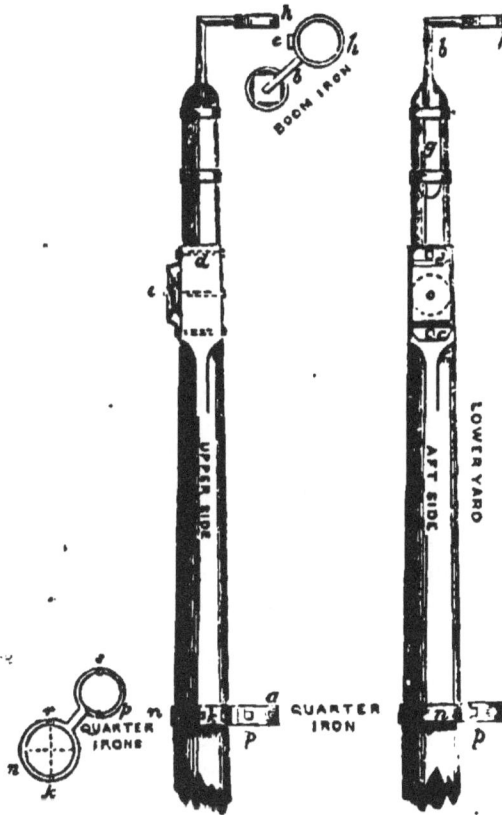

arm, and the crank, *b h*, which projects at right-angles to the strap, having a ring, *h*, connected to it, for the boom: in one side of this ring a horizontal lignum vitæ roller, *e*, is fixed, for the ease of sliding out the booms. The inner irons or quarter-irons, differ in shape to the yard-arm-iron. One part is formed as a clasp hoop to compass the yard, *r k n*, and the ring, *s p*, is separated from it by a shank or chock: the upper part of the ring opens with a hinge, *s*, and key, *p,* for fastening it when the heel of the boom is laid therein.

The boom-irons are fixed on the yards so that a line drawn through the middle of the shanks of each may pass through the angle formed by the foremost eight-square, and foremost upper eight-square, and the centre of any section of the yard. The outer boom-irons let in their thickness at each end of the yard, and are fastened with two hoops driven on tight, with a bolt through between them, and two nails in the ends. The quarter-irons are put on warmed, and the keys driven in.

Boom-irons on the yard-arms of small ships have a straight neck, projecting from straps, with a shoulder in the middle of the neck, and the part without left square. The ring has a shank on the under part, with a square hole that fits the neck, and is there secured by a screw-nut or a spring forelock that goes on the neck next the ring. It is principally large vessels that have quarter-irons on their yards.

ON TOP-SAIL YARDS.

Top-sail yards are of a cylindrical form, tapering from the slings or given diameter towards the ends or yard-arms. The ends are three-sevenths to one-half the given diameter; and the eight squares left on each side of the middle are one-eighth of the given length of the yard, and each side being trimmed sixteen-square, are rounded and planed smooth and fair to their outer ends, except at the places of the sheave-holes, for the top gallant-sheets, which are left square, and cleats or stops raised from the solid, for the reefs of the topsail. These yards have a sheave-hole cut from the upper side its length within each outer end for the reef tackles; and in some merchant ships, sheave-holes are cut for the top gallant-sheets, but a cheek-block fixed on the after side, as shown on the lower-yards, is to be preferred, as sheave-holes weaken the yards.

Finishing of Yards.—Top-sail yards are fitted at the middle with two hoops once the diameter of the yard on each side of the slings, for the topsail ties.

The fore and main top-sail yards commonly have boom-irons at their outer ends, with the outer arm or crank of the iron made to ship and unship. The mizen top-sail yard has no boom-iron, but a ferrule driven on, and an eye driven into the end of the yard.

ON CROSS-JACK YARDS, SPRIT-SAIL YARDS, TOP-GALLANT YARDS, ROYAL YARDS, AND STUDDING-SAIL YARDS.

These yards are at the ends three-sevenths of the diameter at the slings, and are left in the eight squares at the middle one-fourth their length, and each side sixteen squares, then rounded and planed fair and smooth throughout the length, and a ferrule and eye at each end. Stops are formed out of the yard the same as the top-sail yards, and a hoop in the middle for the slings.

ON THE DRIVER OR SPANKER AND MAIN-BOOMS

The given diameter of *driver-booms* is at the middle, and of *main sail-booms* at the sheet or taffrail; their outer ends are two-thirds, and the inner ends three-fourths of the given diameter. They are then lined to the size and rounded all the way through, except for the jaws, *a a*, where it is left square. The jaws are mostly made of oak, and are formed to a half-circle, 1 inch larger than the diameter of the mast, for leather and play: they are in length from 4 to 5 feet from the inner end of the boom, and in depth ¾ to 1 inch less than the diameter of the fore-end. The boom is worked to a tongue to which the jaws are scarphed; they are then formed with a curve inwards, so as to follow in fair to the size of the boom, leaving sufficient strength to the hollow of the jaws, and are then rounded each way the fore-ends. Three or four hoops, *c c c*, are driven over the jaws, and under the third hoop from the fore-end a horizontal bolt is driven through the boom and both parts of the jaws, and one strengthening bolt is also driven about 2½ inches from the hollow; the ends have a nail driven on each side through the jaws into the boom.

Some spanker-booms have no jaws attached to them, but fitted with a goose-neck at the inner end, and a hoop round the mast with an eye to receive it. At the outer end is formed a necking *h*, and has a sheave-hole cut within for the clue of the sail hauling out to the boom-end; one, and sometimes two, hoops with eyes, *d, e*, for the topping-lift; *e*, for the outer, and *d*, for the inner, when it is a long boom.

ON THE MIZEN AND MAIN GAFFS.

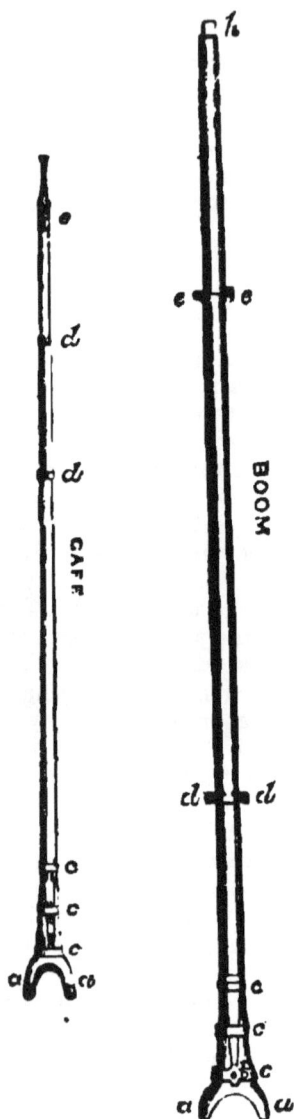

The given diameter of gaffs is at four feet from the inner end, and their lengths are set up from the butt on the upper part; their outer ends are half to five-ninths of the given diameter. They are then lined to the size and rounded, as before. The jaws, *a a*, are made and finished similar to the booms, except that the hollow of the jaws have a great bevelling, that it may be in the direction of the mast, when peaked. An eye-bolt is driven through from the upper side in the direction of the bevelling of the inner end, and clenched underneath, for the throat halyards, and one small eye-bolt with a hook fitted to it driven from the underside, for hooking the neck-earring, and securing the throat downhauler; the jaws are always leathered in the hollow. At the end of the gaff there is a ferrule and eye, up and down, for fixing a small block to for displaying signals. There are likewise hoops with eyes, *d d*, driven on the gaff for the peak halyards.

When the gaff is a " fixed-gaff," there is a goose-neck fitted at the inner-end. *Fore* and *Main-trysail-gaffs* are generally fitted in that manner, which works in the truss-hoop having an eye to receive it; their given diameter is at the inner end, and the outer end is three-fifths of the given diameter. There is no additional length allowed for pole; and a sheave is cut through at the outer end. A ferrule and eye are fitted the same as those of the mizen-gaff.

ON TRY-SAIL-MASTS.

Try-sail-masts are seldom made for ships or barques, but only brigs, termed *snows*, being that they are equipped with a third small mast or try-sail-mast, just abaft the main-mast to carry a sail similar to a ship's spanker. The try-sail-mast is rounded all the way through, and is of an equal diameter the whole length, from one-third to one-half the diameter of the main-mast. The foot of this mast is sometimes fixed in a block of wood, or kind of step, upon the deck; at other times on the boom; and, commonly, steps on a clasp-hoop, with an eye to receive it: the head is secured between the after part of the trestle-trees. Instead of a try-sail-mast, a thick rope called a horse, is mostly used for attaching the fore-leeches of main and fore-try-sails in a ship or a barque.

STUDDING-SAIL-BOOMS.

The given diameter of swinging or lower studding-sail-booms is at the heel and one-third the length, between which they are made parallel, and decrease from thence to two-thirds the given diameter. They are rounded the whole length, and have a necking at the outer-end, and a hole bored through the diameter within; the inner end is fitted with a goose-neck and ferrule.

The top and top-gallant studding-sail-booms have the given diameter at one-third from each end: the ends are two-thirds of the given diameter, and rounded the whole length, and have a hole through the end.

TRESTLE-TREES AND CROSS-TREES FOR LOWER MASTS.

The *trestle-trees* are in length equal to the breadth of the top fore-and-aft; in depth, half the diameter of the mast at the partners; and in breadth or thickness, one-half of the depth. The insides are trimmed straight and out of winding, and the breadth set off parallel to them. The lower sides are snaped from half the depth of the trestle-trees down to one and three-fourths of the depth in length, from the fore end, and one and three-sevenths of the depth from the after end, and rounded to a half-circle at the foremost-ends; the lower edges are chamfered the whole length.

The *cross-trees* are in length the breadth of the top; the breadth,—the breadth of the trestle-trees; and, the depth or thickness, two-thirds of their breadth: they are snaped from half their depth down at the ends to one-fourth of the length of the cross-trees, and their ends rounded off with a sweep; a chamfer is taken off the edges on the under sides and round the ends the length of the snapes.

THE FRAMING OF THE TRESTLE AND CROSS-TREES.

The trestle-trees are placed on horizontal thawts or blocks, and kept apart the breadth of the mast-head athwartships, and square with each other at the ends; the cross-trees are next laid, at their proper stations, athwart the upper sides of the trestle-trees, having the middle of their lengths in the centre between the trestle-trees, and at right angles with them. In letting down the cross-trees, scores are cut in the trestle-trees from three-fourths of an inch to one inch of their depth; a score is also taken out of the under sides of the cross-trees to the breadth, and half the depth on the trestle-trees, to steady them in their places. The cross-trees are then removed, to let up chocks, which are brought on the fore and after sides of the mast between the trestle trees, and of the same depth. It is only in large ships these chocks are fit in the trestle-trees. When the cross-trees are in their places, one saucer-headed bolt is driven through each from the upper side, and screwed-nutted on the under side of the trestle-trees.

TOPS OF MERCHANT-SHIPS.

All tops in the merchant service are made as small and light as possible, in order to reduce the *chafing* of the top-sails against the top rim. The usual length is two-thirds to one-half of their breadth; and the *breadth* of main-top is about equal to one half the moulded breadth of beam; the *breadth* of fore-top nine-tenths of main-top; and *breadth* of mizen-top four-fifths of fore-top. It is not considered necessary to make tops broad for the spread of the top-mast rigging, as it is found now that a good stout *standing backstay* is the main support of the top-mast.

The fore part of tops are of an elliptical form, and a rim of elm or oak board fixed on the cross and trestle-trees, give the external form, when it is either boarded or grated. The framing being formed, iron plates are let in for the puttock-plates: the foremost hole for the puttock-plates is fixed at the centre of the top-mast; the after hole is about six inches from the after part of the top; and the intermediate ones are equally spaced. A hole is cut for the slings at the fore part of the top.

TOP-MAST TRESTLE AND CROSS-TREES.

Tho length of the *trestle-trees* should be governed by the cross-trees and chocks; the breadth $\frac{1}{12}$ of lower trestle-trees of the respective masts; and the depth $\frac{1}{12}$ of the depth of lower trestle-trees. The *cross-trees*, when curved,

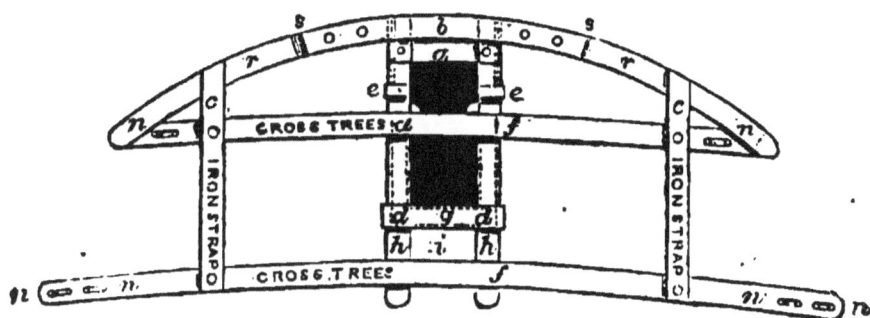

have the proportion thus,—length of after horn, four-sixths of the lower after cross-trees of the respective masts; length of the forward horn, five-sixths of the after one; breadth,—the breadth of the trestle-trees; and, the depth or thickness, four-fifths of the breadth. The horns usually sweep 9 inches in 16 feet.

The trestle-trees are trimmed similar to the lower ones, and snaped at the ends; and the cross-trees have a taper on the under sides to half the depth at the ends, which are rounded, and a hole bored 3 or 4 inches within, for the top-gallant shrouds; instead of a hole, is sometimes a small roller fixed.

The cross-trees and trestle-trees are united together similar to the lower ones, in a frame; and when they are made with a circular piece, *r r*, it is secured to the fore-end of the trestle-trees, *h h*, and to the outer part of the forward horned cross-trees, *f;* a span piece, *b*, unites the two pieces together with four up and down bolts. The short cross piece *a*, between the trestle-trees let down, is bolted to the circular pieces with two small bolts. In fore-top-mast trestle-trees of the navy there is an additional length allowed to them, to give a larger space, *i*, between the top-mast and after-mast cross-tree, for a block for the main top-gallant stay to lead through. The cross-trees and circular pieces *r r*, are connected by an iron strap fixed to them, which is made that it may easily be unfixed.

CHAPTER VI.

Preliminary Remarks.—On Masting Merchant-ships.—Observations on the Diameters and Forms of Masts, Yards, &c.—The greatest Diameter, commonly called the *given Diameter*, where it is situated.—Proportions that the Diameters usually bear to the respective length of the Masts, Yards, &c.—The fractional proportion that the intermediate Diameters bear towards the given Diameter.—Tables of dimensions of Masts and Yards in the Merchant Service.

HAVING extended the practical operations in mast-making over as many pages as the limits of this work will admit, and, it is hoped, as will be sufficient for initiating the young student into the rudiments of this art, a few observations may be added respecting the masting of ships in the Merchant Service.

In the first place it is proper to observe, that in determining the masts and yards for merchant-ships, no general rule seems to be practised, as we find numerous ships in the same trade, of the same dimensions, and (to appearance) of the same stability, with very dissimilar masts and yards. This in some degree may be accounted for, by the masting and sparring of these ships being often regulated by the opinions of those persons who are to command them, sometimes to the fancy of the owner, and at other times to the builder. This matter of tastes has not only caused a great difference in the dimensions of masts and yards for similar ships, but as great a variety in the tautness of the masts and the squareness of the yards; which, go to prove that there is not the proportion maintained of the masts and yards to the length and breadth of the ship, as is the practice in the British Navy.

The rules given by different authors for determining the proportion of masts, yards, &c., for merchant-ships, are of very little use now, for there are so many varieties in the build of these ships,—some very long and narrow, and others both long and broad, that it is impossible to make one rule serve in both cases; and it would require a great number of rules to determine the proportion of masts and yards, the way at present these ships are rigged. It is not intended in this work to give rules, as it is more common now to delineate a plan of the sails, from dimensions of masts and yards of the most approved rig the same size, or thereabouts, and vary them as may be considered necessary for the trade she is to be employed in, and for the number of men that have to work the sails. This is decidedly the best method to adopt for ships of any kind, as it will be seen upon the plan how the masts and yards look in proportion to the size of the vessel; besides, the calculation of the centre of effort of the sails can easily be found, and compared with other ships that have been known to answer well.

ON THE DIAMETERS AND FORMS OF MASTS, YARDS, &C.

In ascertaining what strength and form should be given to the several spars made use of in masting of ships, so that a proper maintenance may be afforded for resisting the strains that masts, yards, &c., are subject to, we must be guided by our knowledge on the effects produced on the different descriptions of timbers, and by what experience and much service have found to be best, rather than by any speculative theory. It is from observations of this sort that the mast-maker has been enabled to form such a judgment respecting the strength and diameters that are requisite for all masts, yards, &c., and to form rules from experience, as to be pretty correct and easy of application.

The largest diameter is that which is *given*, and is taken thus:—at the partners or decks of lower masts; at the cap, in top-masts and top-gallant-masts ; at the bed, of bowsprits ; at the slings or middle, of yards ; at the bowsprit-cap, of jib-booms ; at the middle, of driver-booms ; at the sheet or taffrail, of main-sail-booms ; at four feet from the end, of gaffs ; at one-third from each end of top and top-gallant studding-sail-booms ; and, at one-third from the inner-end of swing or lower studding-sail-booms.

PROPORTIONS THAT THE DIAMETERS USUALLY BEAR TO THE RESPECTIVE LENGTH OF THE MASTS, YARDS, &C.

The diameters given to masts, yards, &c., commonly follow some law of their length, thus :—

Diameters of Masts.

Main and fore-masts, one inch for every three feet of the length.

Main-masts of brigs, one inch to every three feet in length ; and the *fore-mast* nine-tenths of the diameter of the main-mast.

Masts of cutters, three-fourths of an inch in diameter to every three feet in length.

Mizen masts of ships, two-thirds of the diameter of the main-mast.

Main and fore-top-masts, one inch to every three feet in length.

Mizen top-masts, seven-tenths of the diameter of the main-top-mast.

Top-gallant-masts, one inch to every three feet in their length.

Royal-masts, two-thirds the diameter of their top-gallant masts.

Bow-sprits, the same diameter as the main-mast.

Diameters of Yards.

Main and fore-yards, at the slings or middle, seven-tenths to seven-eighths of an inch to every three feet in the length.

Top-sail-yards, five-eighths to seven-eighths of an inch to every three feet in the length.

Top-gallant-yards, six-tenths to five-eighths of an inch to every three feet in the length.

Royal-yards, one-half the diameter of their top-sail yards.

Studding-sail-yards, one inch in diameter to every five feet in the length.

Diameters of Booms.

Studding-sail-booms, one inch to every five feet in the length.

Jib-booms, seven eighths of an inch to every three feet in the length.

Flying-jib-booms, seven-eighths of an inch to every three feet in the length.

Driver-booms, the same as the fore-top-sail-yard.

Gaffs, the same as their booms.

TABLE OF THE FRACTIONAL PROPORTION THAT THE INTERMEDIATE DIAMETERS BEAR TOWARDS THE GIVEN DIAMETER.

Species of Masts, Yards, &c.	Proportions to the given Diameter.					
	Quarters.			Head.		Heel.
	1st.	2nd.	3rd.	Lower Part.	Upper Part.	
Standing-masts . . .	$\frac{63}{64}$	$1\frac{4}{15}$	$\frac{9}{5}$	$\frac{3}{4}$	$\frac{1}{2}$	$\frac{5}{8}$
Topmasts, topgallant-masts, and royal-masts.	$\frac{63}{64}$	$1\frac{4}{15}$	$\frac{9}{5}$	$\frac{9}{15}$	$\frac{9}{11}$	...
Yards	$\frac{39}{64}$	$\frac{7}{8}$	$\frac{7}{10}$	Arms. $\frac{9}{5}$	...	..
Bowsprit	$\frac{63}{64}$	$1\frac{1}{2}$	$\frac{5}{8}$	$\frac{5}{8}$	...	Outer-end. $\frac{4}{7}$
Jib and driver-booms . .	$\frac{42}{64}$	$1\frac{1}{3}$	$\frac{5}{8}$	Ends. $\frac{5}{8}$	...	...
Main-booms . . .	$\frac{44}{64}$	$1\frac{2}{3}$	$\frac{7}{8}$	Fore-end. $\frac{5}{8}$	After-end. $\frac{3}{4}$	Middle. $1\frac{1}{4}$
Gaffs	$\frac{44}{64}$	$1\frac{1}{2}$	$\frac{3}{4}$	$\frac{5}{8}$	...	...
Heeling { Standing-masts	$\frac{5}{8}$ athwartship.			...		$\frac{1}{2}$ fore and aft.
Bowsprit . . }	$\frac{7}{11}$	,,		$\frac{5}{8}$ up and down.		

DIAMETERS OF LOWER MASTS AT THEIR QUARTERS, HEADS, AND HEELS.

Given Diameter.	Quarters.			Heads.		Heels.	Given Diameter.	Quarters.			Heads.		Heels.
	60/61	14/15	6/7	3/4	5/8	6/7		60/61	14/15	6/7	3/4	5/8	6/7
	1st.	2nd.	3rd.	Hounds.	Head.			1st.	2nd.	3rd.	Hounds.	Head.	
Ins.	Ins.	Ins.	Ins.	Ins.	Ins.	Ins.	Ins.	Ins.	Ins.	Ins.	Ins.	Ins.	Ins.
32	31½	29⅞	27⅜	24	20	27⅜	20	19¾	18¾	17⅛	15	12½	17⅛
31½	31	29⅜	27	23⅝	19¾	27	19½	19¼	18¼	16¾	14⅝	12⅛	16¾
31	30½	28⅞	26⅝	23¼	19⅜	26⅝	19	18¾	17¾	16¼	14¼	11⅞	16¼
30½	30	28½	26	22⅞	19⅛	26	18½	18¼	17¼	15⅞	13⅞	11⅝	15⅞
30	29½	28	25¾	22½	18¾	25¾	18	17¾	16¾	15 7/16	13½	11¼	15½
29½	29	27½	25¼	22⅛	18⅜	25¼	17½	17¼	16¼	15	13⅛	11	15
29	28½	27	24⅞	21¾	18⅛	24⅞	17	16¾	15⅞	14⅝	12¾	10⅝	14⅝
28½	28	26½	24⅜	21⅜	17¾	24⅜	16½	16¼	15⅜	14⅛	12⅜	10⅜	14⅛
28	27½	26⅛	24	21	17½	24	16	15¾	14⅞	13¾	12	10	13¾
27½	27	25⅝	23⅝	20⅝	17⅛	23⅝	15½	15¼	14½	13¼	11⅝	9¾	13¼
27	26½	25¼	23⅛	20¼	16⅞	23⅛	15	14¾	14	12⅞	11¼	9⅜	12⅞
26½	26	24¾	22¾	19⅞	16⅝	22¾	14½	14¼	13½	12⅜	10⅞	9⅛	12⅜
26	25½	24¼	22¼	19½	16¼	22¼	14	13¾	13	12	10½	8¾	12
25½	25	23¾	21⅞	19⅛	15⅞	21⅞	13½	13¼	12⅝	11⅝	10⅛	8⅜	11⅝
25	24½	23¼	21⅜	18¾	15⅝	21⅜	13	12¾	12⅛	11⅛	9¾	8⅛	11⅛
24½	24	22⅞	21	18⅜	15¼	21	12½	12¼	11⅝	10¾	9⅜	7⅞	10¾
24	23½	22⅜	20⅝	18	15	20⅝	12	11¾	11¼	10¼	9	7½	10¼
23½	23⅛	21⅞	20⅛	17⅝	14¾	20⅛	11½	11¼	10⅝	9⅞	8⅝	7 3/16	9⅞
23	22⅝	21½	19¾	17¼	14⅜	19¾	11	10¾	10¼	9⅜	8¼	6⅞	9⅜
22½	22⅛	21	19¼	16⅞	14⅛	19¼	10½	10¼	9¾	9	7⅞	6⅝	9
22	21⅝	20½	18⅞	16½	13¾	18⅞	10	9¾	9⅜	8⅝	7½	6¼	8⅝
21½	21⅛	20	18⅜	16⅛	13⅜	18⅜	9½	9¼	8⅞	8⅛	7⅛	5 15/16	8⅛
21	20⅝	19⅝	18	15¾	13⅛	18	9	8⅞	8⅜	7¾	6¾	5⅝	7¾
20½	20⅛	19⅛	17⅝	15⅜	12¾	17⅝							

DIAMETERS OF TOP-MASTS, TOP-GALLANT-MASTS, AND ROYAL-MASTS, AT THEIR QUARTERS AND HEADS.

Diameter at the Cap.	Quarters.			Heads.		Diameter at the Cap.	Quarters.			Heads.	
	$\frac{60}{61}$	$\frac{14}{15}$	$\frac{6}{7}$	$\frac{9}{13}$	$\frac{6}{11}$		$\frac{60}{61}$	$\frac{14}{15}$	$\frac{6}{7}$	$\frac{9}{13}$	$\frac{6}{11}$
	1st.	2nd.	3rd.	Hounds.	Head.		1st.	2nd.	3rd.	Hounds.	Head.
Ins.	Ins.	Ins.	Ins.	Ins.	Ins.	Ins.	Ins.	Ins.	Ins.	Ins.	Ins.
19	18¾	17¾	16¼	13⅛	10⅜	11	10¾	10¼	9½	7⅝	6
18½	18¼	17¼	15¾	12⅞	10⅛	10½	10¼	9¾	9	7¼	5¾
18	17¾	16¾	15½	12½	9⅞	10	9¾	9⅜	8⅝	6$\frac{15}{16}$	5½
17½	17¼	16¼	15	12⅛	9½	9½	9¼	8⅞	8⅛	6$\frac{9}{16}$	5¼
17	16¾	15⅞	14½	11¾	9¼	9	8⅞	8⅜	7¾	6¼	5
16½	16¼	15⅜	14⅛	11⅜	9	8½	8⅜	7⅞	7¼	5$\frac{15}{16}$	4⅝
16	15¾	14⅞	13¾	11$\frac{1}{16}$	8¾	8	7⅞	7$\frac{7}{16}$	6⅞	5½	4⅜
15½	15¼	14½	13¼	10¾	8½	7½	7⅜	7	6⅜	5⅛	4⅛
15	14¾	14	12⅞	10⅜	8¼	7	6⅞	6½	6	4⅞	3⅞
14½	14¼	13½	12½	10	7⅞	6½	6⅜	6$\frac{1}{16}$	5⅝	4½	3½
14	13¾	13	12	9¾	7⅝	6	5⅞	5⅝	5⅛	4⅛	3¼
13½	13¼	12⅝	11⅝	9⅜	7⅜	5½	5⅜	5⅛	4¾	3$\frac{13}{16}$	3
13	12¾	12⅛	11⅛	9	7⅛	5	4⅞	4⅝	4¼	3$\frac{7}{16}$	2¾
12½	12¼	11⅝	10⅝	8⅝	6⅞	4½	4⅜	4⅛	3⅞	3⅛	2½
12	11¾	11⅛	10¼	8$\frac{5}{16}$	6½	4	3⅞	3¾	3⅜	2¾	2¼
11½	11¼	10¾	9⅞	7$\frac{15}{16}$	6¼	3½	3⅜	3¼	3	2⅜	1⅞

DIAMETERS OF THE BOWSPRITS, AT THEIR QUARTERS, ETC.

Diameter at the Bed.	Quarters. 1st. 60/61	Quarters. 2nd. 11/12	Quarters. 3rd. 4/5	2/3 Outer End.	6/7 Heels.	Diameter at the Bed.	Quarters. 1st. 60/61	Quarters. 2nd. 11/12	Quarters. 3rd. 4/5	2/3 Outer End.	6/7 Heels.
Ins.	Ins.	Ins.	Ins.	Ins.	Ins.	Ins.	Ins.	Ins.	Ins.	Ins.	Ins.
31	30½	28⅜	24¾	20⅝	26⅝	21½	21⅛	19¾	17⅛	14 5/16	18⅜
30½	30	28	24⅜	20⅜	26	21	20⅝	19¼	16¾	14	18
30	29½	27½	24	20	25¾	20½	20⅛	18⅞	16⅜	13 11/16	17⅝
29½	29	27	23⅝	19⅝	25¼	20	19⅝	18⅜	16	13 5/16	17⅛
29	28½	26⅝	23¼	19⅜	24⅞	19½	19⅛	17⅞	15⅝	13	16¾
28½	28	26	22¾	19	24⅜	19	18¾	17⅜	15¼	12 11/16	16¼
28	27½	25⅝	22⅜	18⅝	24	18½	18⅛	17	14¾	12⅜	15⅞
27½	27	25¼	22	18⅜	23½	18	17¾	16½	14⅜	12	15⅜
27	26½	24¾	21⅝	18	23⅛	17½	17⅛	16	14	11 11/16	15
26½	26	24¼	21¼	17⅝	22¾	17	16¾	15⅝	13⅝	11¼	14½
26	25½	23⅞	20¾	17 5/16	22¼	16½	16⅛	15⅛	13⅛	11	14⅛
25½	25	23⅜	20⅜	17	21⅞	16	15¾	14⅝	12¾	10⅝	13¾
25	24½	22⅞	20	16⅝	21⅜	15½	15¼	14⅛	12⅜	10¼	13¼
24½	24	22½	19⅝	16 5/16	21	15	14¾	13⅞	12	10	12⅞
24	23½	22	19¼	16	20½	14½	14¼	13¼	11⅝	9¾	12⅜
23½	23⅛	21½	18¾	15⅝	20⅛	14	13¾	12⅞	11⅛	9⅜	12
23	22⅝	21⅛	18⅜	15 5/16	19¾	13½	13¼	12⅜	10¾	9	11½
22½	22⅛	20⅝	18	15	19¼	13	12¾	11⅞	10⅜	8¾	11⅛
22	21⅝	20⅛	17⅝	14⅝	18⅞	12½	12¼	11½	10	8¼	10⅝

DIAMETERS OF YARDS AT THEIR QUARTERS.

Diameter at the Slings.	Quarters.				Diameter at the Slings.	Quarters.			
	30/31	7/8	7/10	3/7		30/31	7/8	7/10	3/7
	1st.	2nd.	3rd.	Arm.		1st.	2nd.	3rd.	Arm.
Ins.	Ins.	Ins.	Ins.	Ins.	Ins.	Ins.	Ins.	Ins.	Ins.
22	21¼	19¼	15⅜	9 1/16	12½	12 1/16	10⅞	8¾	5⅜
21½	20¾	18⅞	15	9¼	12	11⅝	10½	8⅜	5⅛
21	20¼	18⅜	14¾	9	11½	11 1/16	10	8	4 15/16
20½	19⅞	17 15/16	14⅜	8¾	11	10⅝	9⅝	7¾	4¾
20	19⅜	17½	14	8 9/16	10½	10 1/16	9⅛	7⅜	4½
19½	18⅞	17	13⅝	8⅜	10	9⅝	8¾	7	4¼
19	18⅜	16⅝	13¼	8⅛	9½	9⅛	8¼	6⅝	4 1/16
18½	17 15/16	16⅛	13	8	9	8¾	7⅞	6¼	3⅞
18	17⅜	15¾	12⅝	7¾	8½	8¼	7⅜	5 15/16	3⅝
17½	17	15¼	12¼	7½	8	7¾	7	5⅝	3⅜
17	16½	14⅞	11⅞	7¼	7½	7¼	6½	5¼	3¼
16½	16	14⅜	11⅝	7 1/16	7	6¾	6⅛	4 15/16	3
16	15½	14	11 3/16	6⅞	6½	6¼	5¾	4½	2¾
15½	15	13 9/16	10⅞	6⅝	6	5⅞	5¼	4 3/16	2 9/16
15	14½	13⅛	10½	6⅜	5½	5 5/16	4⅞	3⅞	2⅜
14½	14	12¾	10⅛	6¼	5	4⅞	4⅜	3½	2⅛
14	13½	12¼	9¾	6	4½	4⅜	3 15/16	3⅛	1 15/16
13½	13	11⅞	9½	5¾	4	3⅞	3½	2 13/16	1¾
13	12½	11⅜	9 1/16	5 9/16	3½	3⅜	3 1/16	2 7/16	1½

DIAMETERS OF DRIVER-BOOMS, JIB-BOOMS, ETC., AT THEIR QUARTERS.

Driver-booms, Jib-booms, &c.					Driver-booms, Jib-booms, &c.				
Given Diameters.	Quarters.			Ends.	Given Diameters.	Quarters.			Ends.
	40/41	11/12	5/6	2/3		40/41	11/12	5/6	2/3
	1st.	2nd.	3rd.			1st.	2nd.	3rd.	
Ins.	Ins.	Ins.	Ins.	Ins.	Ins.	Ins.	Ins.	Ins.	Ins.
16	15⅝	14⅝	13⅜	10⅝	9½	9¼	8¾	7 15/16	6⅜
15½	15 1/16	14¼	13	10¼	9	8¾	8¼	7½	6
15	14⅝	13¾	12½	10	8½	8¼	7¾	7 1/16	5⅝
14½	14 1/16	13¼	12	9⅝	8	7⅞	7⅜	6⅝	5⅜
14	13⅝	12⅞	11⅝	9⅜	7½	7⅜	6⅞	6¼	5
13½	13⅛	12⅜	11¼	9	7	6⅞	6⅜	5⅞	4⅝
13	12⅝	11 15/16	10⅞	8⅝	6½	6⅜	6	5 7/16	4⅜
12½	12⅛	11½	10 7/16	8¼	6	5⅞	5½	5	4
12	11¾	11	10	8	5½	5⅜	5 1/16	4⅝	3 11/16
11½	11¼	10½	9⅝	7⅝	5	4⅞	4⅝	4⅛	3⅜
11	10¾	10 1/16	9⅛	7⅜	4½	4⅜	4⅛	3¾	3
10½	10¼	9⅝	8¾	7	4	3⅞	3⅝	3⅜	2⅝
10	9¾	9⅛	8⅜	6⅝	3½	3 7/16	3¼	2 15/16	2⅜

DIAMETERS OF MAIN-BOOMS AND GAFFS, AT THEIR QUARTERS.

Main-booms.							Gaffs.				
Given Diameter.	Quarters.			Fore-end.	Middle.	After-end.	Given Diameter.	Quarters.			End.
	40/41	11/12	7/8					40/41	11/12	4/5	5/9
	1st.	2nd.	3rd.	2/3	11/12	3/4		1st.	2nd.	3rd.	
Ins.	Ins.	Ins.	Ins.	Ins.	Ins.	Ins.	Ins.	Ins.	Ins.	Ins.	Ins.
16	15⅝	14¾	14	10⅝	14⅝	12	12	11¾	11	9⅝	6⅝
15½	15⅛	14¼	13½	10¼	14¼	11⅝	11½	11⅛	10½	9⅛	6⅜
15	14⅝	13⅞	13⅛	10	13¾	11¼	11	10¾	10 1/16	8¾	6⅛
14½	14 1/16	13⅜	12⅝	9¾	13¼	10⅞	10½	10¼	9⅝	8⅜	5⅞
14	13¾	13	12¼	9⅜	12¾	10½	10	.9¾	9⅛	8	5 9/16
13½	13⅛	12⅜	11¾	9	12⅜	10⅛	9½	9¼	8¾	7 9/16	5¼
13	12¾	12	11⅜	8¾	11 15/16	9¾	9	8¾	8¼	7¼	5
12½	12⅛	11 7/16	10⅞	8¼	11½	9⅜	8½	8¼	7¾	6¾	4¾
12	11¾	11 1/16	10½	8	11	9	8	7⅞	7⅜	6⅜	4 7/16
11½	11⅛	10 9/16	10	7⅝	10½	8⅝	7½	7⅜	6⅞	6	4⅛
11	10¾	10⅛	9⅝	7⅜	10 1/16	8¼	7	6⅞	6⅜	5 9/16	3⅞
10½	10¼	9 11/16	9⅛	7	9⅝	7⅞	6½	6⅜	6	5 3/16	3⅝
10	9¾	9¼	8¾	6⅝	9⅛	7½	6	5⅞	5½	4¾	3 5/16
9½	9¼	8⅞	8¼	6¼	8¾	7⅛					
9	8¾	8 5/16	7⅞	6	8¼	6¾					
8½	8¼	7 13/16	7⅜	5⅝	7¾	6⅜					
8	7⅞	7⅜	7	5⅜	7⅜	6					

DIMENSIONS OF MASTS AND YARDS IN THE MERCHANT SERVICE.

Names of the Masts and Yards.	Ship 1568 tons. Dimensions of Ship: Length, 212 ft. 6 in.; Breadth, 42 ft. 6 in.						Ship 1415 tons. Dimensions of Ship: Length, 190 ft.; Breadth, 41 ft.						Ship 1330 tons. Dimensions of Ship: Length, 176 ft.; Breadth, 40 ft.					
	Masts or Booms.			Yards.			Masts or Booms.			Yards.			Masts or Booms.			Yards.		
	Extreme Length.	Headed Length.	Diameter.	Extreme Length.	Arm.	Diameter.	Extreme Length.	Headed Length.	Diameter.	Extreme Length.	Arm.	Diameter.	Extreme Length.	Headed Length.	Diameter.	Extreme Length.	Arm.	Diameter.
	Ft. In.	Ft. In.	Ins.	Ft. In.	Ft. In.	Ins.	Ft. In.	Ft. In.	Ins.	Ft. In.	Ft. In.	Ins.	Ft. In.	Ft. In.	Ins.	Ft. In.	Ft. In.	Ins.
Main-mast and yard	101 0	16 0	32	89 6	4 9	22	96 0	15 6	30	84 0	4 0	22	98 0	15 0	30	82 6	4 3	20½
Top-mast and yard	60 0	8 6	19	70 0	5 3	16	58 0	8 6	19	68 0	5 0	16	55 3	8 3	17	64 0	4 9	15
Topgallant-mast and yard	34 0	..	14	50 0	3 0	11½	32 0	..	13	46 0	2 6	11	29 0	..	10	45 8	2 8	10
Royal-mast and yard	22 6	..	8½	39 8	2 2	8½	21 0	..	8½	36 0	1 9	8½	20 9	..	6½	34 6	1 8	7½
Fore-mast and yard	94 0	15 0	32	82 6	4 9	22	91 0	15 0	30	76 0	4 0	21½	80 0	14 0	29½	73 0	4 0	17½
Top-mast and yard	56 0	8 0	18½	66 0	5 0	16	53 0	8 0	18½	62 0	5 0	15½	51 0	8 0	16½	58 0	4 9	12½
Topgallant-mast and yard	27 6	..	12¾	46 0	3 9	11½	28 0	..	13	42 0	2 6	11	26 6	..	9	41 6	2 6	9½
Royal-mast and yard	21 0	..	8	35 6	[illegible] 0	8½	19 6	..	8	32 0	1 9	6	18 9	..	7	30 3	1 6	6
Mizen-mast and cross-jack yard	80 0	10 10	23½	72 0	6 0	15	85 0	11 9	20	68 0	6 0	14½	69 0	10 4	19	63 9	6 0	13¾
Top-mast and yard	45 6	0 9	15	65 0	4 2	13	42 0	6 6	14½	50 0	3 6	12	40 9	6 4	12½	46 9	3 3	10
Topgallant-mast and yard	24 0	..	9	42 0	2 6	10	22 0	..	10	38 0	2 0	9½	21 0	..	7	34 3	1 10	7¾
Royal-mast and yard	19 6	..	7	34 0	2 0	6½	16 0	..	5½	28 0	1 6	5⅝	15 0	..	5	26 0	1 4	5½
Bowsprit	61 0	..	31	..	..	..	56 0	..	30	..	..	..	54 6	..	29	..	..	..
								End.										
Jib-boom	50 0	..	15	..	..	..	48 0	2 6	14½	..	..	..	48 0	..	14½	..	..	..
Flying-jib-boom	50 0	..	11	..	..	..	52 0	1 0	10½	..	..	..	51 0	..	10½	..	..	..
Driver-boom	..	..	..	..	..	..	64 0	1 0	13½	..	[illegible]	..	60 0	..	13	..	..	..
Gaff	44 0	3 0	11½	..	..	..	47 0	6 6	11	..	..	..	45 6	..	11	..	..	..

DIMENSIONS OF MASTS AND YARDS IN THE MERCHANT SERVICE.

Names of the Masts and Yards.	Ship 1260 tons. Dimensions of Ship: Length, 165 ft.; Breadth, 37 feet 6 in. — Masts or Booms. Extreme Length. Ft. In.	Headed Length. Ft. In.	Diameter. Ins.	Yards. Extreme Length. Ft. In.	Arm. Ft. In.	Diameter. Ins.	Ship 1160 tons. Dimensions of Ship: Length, 152 ft.; Breadth, 36 ft. 6 in. — Masts or Booms. Extreme Length. Ft. In.	Headed Length. Ft. In.	Diameter. Ins.	Yards. Extreme Length. Ft. In.	Arm. Ft. In.	Diameter. Ins.	Ship 1000 tons. Dimensions of Ship: Length, 150 ft.; Breadth, 35 ft. 6 in. — Masts or Booms. Extreme Length. Ft. In.	Headed Length. Ft. In.	Diameter. Ins.	Yards. Extreme Length. Ft. In.	Arm. Ft. In.	Diameter. Ins.
Main-mast and yard	80 6	14 6	29½	79 6	4 3	19	86 0	13 10	27½	76 0	4 3	18	83 0	18 0	26½	74 0	4 0	17½
Top-mast and yard	53 6	8 0	16½	62 0	4 9	14	51 6	8 0	16	60 0	4 9	14	50 0	7 6	16	59 0	4 9	13¾
Topgallant-mast and yard	28 3	..	10	44 0	2 6	9	27 6	..	10	42 6	2 6	9	28 6	..	10	41 6	2 9	9
Royal-mast and yard	19 9	..	7	33 6	1 6	7	19 0	..	7	32 0	1 6	6½	20 6	..	7	30 6	1 9	6½
Fore-mast and yard	83 0	13 6	27	70 6	4 0	16½	80 6	13 0	26½	68 0	4 0	16	78 0	12 6	26	66 0	[illegible]	16
Top-mast and yard	49 3	7 9	15½	56 6	4 6	12	47 6	7 6	15½	55 0	4 6	11½	46 0	7 0	15½	54 0	[illegible]	11½
Topgallant-mast and yard	26 9	..	9	40 0	2 4	8	25 0	..	9	38 6	2 4	8	25 6	..	9	38 0	2 6	8
Royal-mast and yard	18 0	..	6	29 8	1 6	6	17 6	..	6	28 0	1 6	6	17 6	..	6	27 0	1 6	6
Mizen-mast and yard	65 6	10 6	19	62 0	6 0	13¼	63 0	10 6	16	60 9	6 0	13	60 0	10 6	17	60 0	6 0	13
Top-mast and yard	39 3	6 0	12	45 4	3 0	9½	38 6	6 0	12	44 0	3 0	9½	38 0	6 0	12	43 0	3 0	9½
Topgallant-mast and yard	20 6	..	7	33 0	1 9	7	20 0	..	7	31 6	1 9	6½	20 0	..	7	30 6	2 0	6½
Royal-mast and yard	14 6	..	5	24 9	1 3	5	14 0	..	5	23 6	1 6	5	14 0	..	5	22 6	1 6	5
Bowsprit	[illegible] 0	..	..	..	..	..	52 6	..	..	..	..	..	52 0	..	27	..	..	..
Jib-boom	46 0	End. 2 0	..	..	..	..	46 0	End. 2 0	..	..	..	..	42 6	End. 2 0	13½	..	..	..
Flying-jib-boom	50 0	1 0	..	..	..	..	49 6	1 0	..	..	..	..	45 0	1 0	9	..	..	..
Driver-boom	58 6	1 0	..	..	..	..	52 0	1 0	..	..	..	..	50 0	1 0	8½	..	..	..
Gaff	43 6	5 9	..	..	..	..	43 0	5 6	..	..	..	..	40 0	4 6	10½	..	..	..

DIMENSIONS OF MASTS AND YARDS IN THE MERCHANT SERVICE.

Names of the Masts and Yards.	Ship of 960 tons. Dimensions of Ship: Length, 156 ft.; Breadth, 35 ft.						Ship of 800 tons. Dimensions of Ship: Length, 152 ft.; Breadth, 34 ft. 6 in.						Ship of 730 tons. Dimensions of Ship: Length, 129 ft. 6 in.; Breadth, 32 ft. 9 in.					
	Masts or Booms.			Yards.			Masts or Booms.			Yards.			Masts or Booms.			Yards.		
	Extreme Length.	Headed Length.	Diameter.	Extreme Length.	Arm.	Diameter.	Extreme Length.	Headed Length.	Diameter.	Extreme Length.	Arm.	Diameter.	Extreme Length.	Headed Length.	Diameter.	Extreme Length.	Arm.	Diameter.
	Ft. In.	Ft. In.	Ins.	Ft. In.	Ft. In.	Ins.	Ft. In.	Ft. In.	Ins.	Ft. In.	Ft. In.	Ins.	Ft. In.	Ft. In.	Ins.	Ft. In.	Ft. In.	Ins.
Main-mast and yard . .	82 0	13 0	26½	73 6	4 0	20	80 0	18 0	26	72 0	4 0	16	77 0	12 6	22¾	63 0	4 0	14¾
Top-mast and yard . .	49 6	7 8	16	53 0	4 6	14	48 6	7 6	15½	55 0	4 6	12½	47 0	6 9	14½	50 0	4 0	11
Topgallant-mast and yard .	25 9	..	7	43 0	2 8	9¾	24 0	..	8¾	39 0	2 6	8½	26 0	..	7	38 6	2 6	8
Royal-mast and yard . .	17 6	..	6	30 0	1 8	6	16 6	..	6¾	27 6	1 9	6½	20 0	..	6¾	25 0	1 8	5
Fore-mast and yard . .	77 0	13 0	26	65 6	3 0	16½	75 0	12 6	25½	64 0	3 9	15½	71 0	11 6	22½	58 0	3 6	13¾
Top-mast and yard . .	45 6	7 2	15½	52 6	4 0	12	44 0	7 0	15	49 6	4 3	12	43 6	6 0	14	46 0	3 9	10½
Topgallant-mast and yard .	23 6	..	7	37 6	2 6	7¾	22 0	..	8½	35 0	2 3	8¼	24 0	..	7	36 0	2 8	8
Royal-mast and yard . .	16 0	..	5½	26 0	1 6	5¼	15 0	..	6½	24 6	1 9	6¼	19 0	..	6¾	23 0	1 2	5
Mizen-mast and yard . .	59 0	10 0	20	59 0	6 0	13	57 6	9 6	17	56 0	6 0	11	68 0	9 0	15	50 0	4 6	10½
Top-mast and yard . .	37 0	5 8	12	42 0	3 6	9	36 6	5 9	11½	38 0	3 0	8½	36 0	5 3	10	37 0	3 0	9
Topgallant-mast and yard .	19 6	..	6½	30 0	2 0	6¼	18 8	..	6½	27 0	1 6	6	20 0	..	6½	27 0	2 0	6
Royal-mast and yard . .	13 0	..	4½	21 6	1 6	4½	12 0	..	5½	19 0	1 0	5	16 0	..	5½	21 0	1 3	4¾
Bowsprit	50 0	..	26½	..	..	..	48 0	..	25	..	..	..	47 0	..	23	..	..	..
		End.						End.						End.				
Jib-boom	45 0	2 0	14	..	..	..	44 0	2 0	11¾	..	..	..	40 0	1 6	11	..	..	..
Flying jib-boom	49 0	..	9	..	..	..	48 0	..	8½	..	..	..	41 0	..	7¾	..	..	..
Driver-boom or out-rigger .	..	..	..	..	..	..	..	..	..	..	..	..	16 0	..	..	..	..	..
Gaff	41 6	6 0	9	..	..	..	40 0	6 0	8½	..	..	..	38 6	4 6	8½	..	..	..

DIMENSIONS OF MASTS AND YARDS IN THE MERCHANT SERVICE.

Names of the Masts and Yards.	Ship of 600 tons. Dimensions of Ship: Length, 125 ft.; Breadth, 28 ft. 6 in.						Barque of 623 tons. Dimensions of Ship: Length, 120 ft.; Breadth, 32 ft.						Ship of 508 tons. Dimensions of Ship: Length, 118 ft. 6 in.; Breadth, 29 ft. 8 in.					
	Masts or Booms.			Yards.			Masts or Booms.			Yards.			Masts or Booms.			Yards.		
	Extreme Length.	Headed Length.	Diameter.	Extreme Length.	Arm.	Diameter.	Extreme Length.	Headed Length.	Diameter.	Extreme Length.	Arm.	Diameter.	Extreme Length.	Headed Length.	Diameter.	Extreme Length.	Arm.	Diameter.
	Ft. In.	Ft. In.	Ins.	Ft. In.	Ft. In.	Ins.	Ft. In.	Ft. In.	Ins.	Ft. In.	Ft. In.	Ins.	Ft. In.	Ft. In.	Ins.	Ft. In.	Ft. In.	Ins.
Main-mast and yard	69 0	10 6	22½	60 0	3 0	14½	71 9	10 6	22½	59 6	4 0	15	09 0	10 0	22½	57 0	3 3	13½
Top-mast and yard	41 0	6 6	13½	48 0	3 0	12	41 3	6 6	13¾	47 0	3 0	12	41 6	0 6	13½	46 0	3 0	10
Topgallant-mast and yard	22 0	..	8	36 0	2 0	8	22 0	..	8	35 0	2 0	7½	21 6	..	7¾	33 0	2 0	7½
Royal-mast and yard	16 6	..	5½	24 6	1 3	6½	14 0	..	5	23 0	1 3	6½	16 0	..	5½	21 6	1 3	6
Fore-mast and yard	66 6	0 6	22	57 0	3 0	14	63 6	10 0	22½	57 0	4 0	14½	64 0	9 0	21½	53 0	3 8	13
Top-mast and yard	39 0	6 0	13	45 0	3 6	11	40 5	5 9	13½	45 6	3 9	11½	38 0	5 0	13	41 0	3 3	10
Topgallant-mast and yard	21 0	..	8	33 0	1 0	7½	23 0	..	8	32 0	1 0	7½	20 0	..	7½	30 0	1 9	7
Royal-mast and yard	15 0	..	5½	23 0	1 0	0	13 6	..	4½	23 0	1 3	6	15 0	..	5½	20 0	1 0	4½
Mizen-mast and yard	01 0	8 0	17	47 0	5 0	10	01 6	8 0	17	..	..	..	59 0	8 0	16	44 0	4 0	9½
Top-mast and yard	31 0	4 0	9	34 0	3 0	8	34 0	Polo. 13 0	9¾	..	..	..	39 0	4 3	0	31 0	3 0	7½
Topgallant-mast and yard	17 0	..	6½	25 0	1 6	5½	..	..	..	..	..	..	10 0	..	6	22 0	1 0	5
Royal-mast and yard	10 0	..	4	10 0	1 0	4½	..	..	..	..	..	..	10 0	..	4	10 0	1 0	4
Bowsprit	45 0	..	23	..	..	..	44 6	..	23	..	..	..	41 0	..	22	..	..	..
Jib-boom	40 0	..	11½	..	..	..	40 0	End. 1 10	11½	..	..	..	40 0	End. 1 8	11½	..	..	..
Flying-jibboom	42 0	..	8¼	..	..	..	40 0	1 4	8¼	..	..	..	41 0	1 0	8¼	..	..	..
Driver-boom	42 0	..	12	..	..	..	41 0	1 0	11	..	..	..	..	..	..	..	..	..
Gaff	32 0	4 0	8	..	..	..	32 0	3 6	8	..	..	..	32 0	5 0	7½	..	..	..

DIMENSIONS OF MASTS AND YARDS IN THE MERCHANT SERVICE.

Names of the Masts and Yards.	Barque of 460 tons. Dimensions of Ship: Length, 115 ft.; Breadth, 28 ft. 6 in. — Masts or Booms. Extreme Length.	Masts or Booms. Headed Length.	Masts or Booms. Diameter.	Yards. Extreme Length.	Yards. Arm.	Yards. Diameter.
	Ft. In.	Ft. In.	Ins.	Ft. In.	Ft. In.	Ins.
Main-mast and yard	63 0	10 0	20½	55 6	8 6	13¼
Top-mast and yard	35 9	5 9	13¾	45 0	8 0	11
Topgallant-mast and yard	17 6	..	8	33 0	2 3	8
Royal-mast and yard	10 6	..	4	24 0	1 6	5¾
Fore-mast and yard	61 6	10 0	20½	55 6	8 6	13¼
Top-mast and yard	35 9	5 9	13¾	45 0	8 0	11
Topgallant-mast and yard	17 6	..	8	33 0	2 8	8
Royal-mast and yard	10 6	..	4	24 0	1 6	5¾
Mizen-mast	61 0	9 6	15½	..	..	..
Top-mast	32 0	..	8½	..	..	..
Pole	10 6	..	5	..	..	..
Bowsprit	35 0	..	22	..	..	..
		End.				
Jib-boom	35 0	..	11¾	..	..	..
Flying-jibboom	9 0	..	..	..	..	..
Mizen-boom	35 0	..	8	..	..	..
Gaff	31 0	4 6	7½	..	..	..
Main and Fore-gaffs	22 0	..	9	..	..	..

Names of the Masts and Yards.	Barque of 430 tons. Dimensions of Ship: Length, 114 ft.; Breadth, 28 ft. — Masts or Booms. Extreme Length.	Masts or Booms. Headed Length.	Masts or Booms. Diameter.	Yards. Extreme Length.	Yards. Arm.	Yards. Diameter.
	Ft. In.	Ft. In.	Ins.	Ft. In.	Ft. In.	Ins.
Main-mast and yard	61 0	9 9	20	53 6	3 0	13
Top-mast and yard	35 6	5 6	13¼	44 0	3 0	10½
Topgallant-mast and yard	17 6	..	8	32 0	2 0	7½
Royal-mast and yard	10 6	..	4	23 6	1 6	5½
Fore-mast and yard	50 6	9 9	20	53 6	8 0	13
Top-mast and yard	35 6	5 6	13¼	44 0	8 0	10½
Topgallant-mast and yard	17 6	..	8	32 0	2 0	7½
Royal-mast and yard	10 6	..	4	23. 6	1 6	5½
Mizen-mast	61 0	8 0	15½	..	..	..
Top-mast	31 0	..	8	..	..	..
Pole	10 0	..	5	..	..	..
Bowsprit	35 0	..	22	..	..	..
		End.				
Jib-boom	35 0	..	11¾	..	..	..
Flying-jibboom	9 0	..	..	..	..	..
Mizen-boom	34 0	..	8	..	..	..
Gaff	30 0	4 0	7½	..	..	..
Main and Fore-gaffs	20 0	..	8½	..	..	..

Names of the Masts and Yards.	Barque of 430 tons. Dimensions of Ship: Length, 113 ft.; Breadth, 28 ft. — Masts or Booms. Extreme Length.	Masts or Booms. Headed Length.	Masts or Booms. Diameter.	Yards. Extreme Length.	Yards. Arm.	Yards. Diameter.
	Ft. In.	Ft. In.	Ins.	Ft. In.	Ft. In.	Ins.
Main-mast and yard	63 0	9 8	20½	52 0	3 5	13
Top-mast and yard	35 0	5 0	13¼	48 0	3 0	10½
Topgallant-mast and yard	17 6	..	8	33 0	2 4	7½
Royal-mast and yard	10 6	..	4	23 0	1 9	5½
Fore-mast and yard	60 6	9 8	20	52 0	8 8	13
Top-mast and yard	35 0	5 0	13¼	43 0	8 0	10½
Topgallant-mast and yard	17 6	..	8	33 0	2 4	7½
Royal-mast and yard	10 6	..	4	23 0	1 9	5½
Mizen-mast	63 0	8 0	16½	..	..	..
Top-mast	35 0	..	9¾	..	..	..
Pole	9 8	..	..	..	..	..
Bowsprit	34 9	..	21	..	..	..
Jib-boom	38 0	..	12	..	..	..
Flying-jibboom	39 0	..	8	..	..	..
Mizen-boom	35 0	..	9	..	..	..
Gaff	30 0	..	7½	..	..	..
Main and Fore-gaffs	21 0	..	8¾	..	..	..

DIMENSIONS OF MASTS AND YARDS IN THE MERCHANT SERVICE.

Barque of 410 tons. Dimensions of Ship: Length, 104 ft.; Breadth, 27 ft. 9 in.

Names of the Masts and Yards.	Masts or Booms. Extreme Length. Ft. In.	Masts or Booms. Headed Length. Ft. In.	Masts or Booms. Diameter. Ins.	Yards. Extreme Length. Ft. In.	Yards. Arm. Ft. In.	Yards. Diameter. Ins.
Main-mast and yard	66 0	9 6	21	55 0	3 0	13½
Top-mast and yard	36 0	5 3	13¼	44 0	3 6	11
Topgallant-mast and yard	19 0	..	8	31 0	2 0	7½
Royal-mast and yard	12 0	..	4½	21 0	1 6	5
Fore-mast and yard	62 0	9 0	20½	53 0	3 0	13
Top-mast and yard	36 0	5 3	13¼	42 0	3 0	11
Topgallant-mast and yard	18 0	..	8	29 6	1 9	7
Royal-mast and yard	11 0	..	4½	20 0	1 3	4½
Mizen-mast	63 0	7 9	17	..	..	..
Top-mast	31 10	..	8½	..	..	..
Pole	12 0	..	5½	..	..	..
Bowsprit	36 0	..	22	..	..	..
Jib-boom	33 0	End. 1 0	11½	..	..	..
Flying-jibboom	39 0	1 6	8	..	..	..
Mizen-boom	34 0	..	8	..	..	..
Gaff	30 0	5 0	7½	..	..	..
Main-gaff	21 0	0 6	8¾	..	..	..

Barque of 400 tons. Dimensions of Ship: Length, 101 ft. 6 in.; Breadth, 28 ft. 6 in.

Names of the Masts and Yards.	Masts or Booms. Extreme Length. Ft. In.	Masts or Booms. Headed Length. Ft. In.	Masts or Booms. Diameter. Ins.	Yards. Extreme Length. Ft. In.	Yards. Arm. Ft. In.	Yards. Diameter. Ins.
Main-mast and yard	68 0	10 9	21	54 0	3 6	14
Top-mast and yard	38 0	6 0	12½	41 6	3 9	10½
Topgallant-mast and yard	18 6	..	7½	28 0	2 6	7
Royal-mast and yard	11 6	Pole. 6 0	4½	19 0	1 6	5¼
Fore-mast and yard	65 0	10 6	20½	51 0	3 6	12¾
Top-mast and yard	36 0	5 6	12	39 0	3 6	10
Topgallant-mast and yard	17 6	..	7½	26 6	2 6	7
Royal-mast and yard	10 6	Pole. 5 6	4½	17 6	1 6	5¼
Mizen-mast	62 0	9 0	14	..	..	..
Top-mast	40 6	..	8	..	..	..
Pole	9 0	5 6	6	..	..	..
Bowsprit	42 0	..	22	..	..	..
Jib-boom	38 0	..	10½	..	..	..
Flying-jibboom	40 0	..	7	..	..	..
Mizen-boom	35 0	..	8¼	..	..	..
Gaff	29 0	4 0	7½	..	..	..
Main-gaff	22 0	..	8¾	..	..	..

Barque of 480 tons. Dimensions of Ship: Length, 117 ft.; Breadth, 29 ft.

Names of the Masts and Yards.	Masts or Booms. Extreme Length. Ft. In.	Masts or Booms. Headed Length. Ft. In.	Masts or Booms. Diameter. Ins.	Yards. Extreme Length. Ft. In.	Yards. Arm. Ft. In.	Yards. Diameter. Ins.
Main-mast and yard	66 0	10 0	21½	57 0	3 8	13½
Top-mast and yard	39 0	6 3	12½	45 0	3 0	11
Topgallant-mast and yard	20 0	..	7¾	34 0	2 6	8
Royal-mast and yard	11 0	..	4	26 0	2 0	5¾
Fore-mast and yard	63 6	10 0	20½	57 0	3 8	13½
Top-mast and yard	39 0	6 3	12½	45 0	3 0	11
Topgallant-mast and yard	20 0	..	7¾	34 0	2 6	8
Royal-mast and yard	11 0	..	4	25 0	2 0	5¾
Mizen-mast	64 6	9 0	17½	..	..	..
Top-mast	35 0	11 0	9¾	..	..	..
Pole	4 0	..	..	..	..	..
Bowsprit	38 0	..	21	..	..	..
Jib-boom	35 0	..	11½	..	..	..
Flying-jibboom	39 0	..	8	..	..	..
Mizen-boom	39 0	..	9	..	..	..
Gaff	36 0	..	8	..	..	..
Main-gaff	21 0	..	8½	..	..	..

DIMENSIONS OF MASTS AND YARDS IN THE MERCHANT SERVICE.

Names of the Masts and Yards.	Barque of 360 tons. Dimensions of Ship: Length, 101 ft.; Breadth, 27 ft. 9 in.						Barque of 320 tons. Dimensions of Ship: Length, 100 ft. 6 in.; Breadth, 27 ft. 6 in.						Barque of 300 tons. Dimensions of Ship: Length, 100 ft.; Breadth, 27 ft.					
	Masts or Booms.			Yards.			Masts or Booms.			Yards.			Masts or Booms.			Yards.		
	Extreme Length.	Headed Length.	Diameter.	Extreme Length.	Arm.	Diameter.	Extreme Length.	Headed Length.	Diameter.	Extreme Length.	Arm.	Diameter.	Extreme Length.	Headed Length.	Diameter.	Extreme Length.	Arm.	Diameter.
	Ft. In.	Ft. In.	Ins.	Ft. In.	Ft. In.	Ins.	Ft. In.	Ft. In.	Ins.	Ft. In.	Ft. In.	Ins.	Ft. In.	Ft. In.	Ins.	Ft. In.	Ft. In.	Ins.
Main-mast and yard . .	64 6	9 6	20¾	53 0	3 3	13½	63 0	9 6	19	51 6	3 0	12½	62 0	9 0	18½	50 0	3 0	12
Top-mast and yard . .	35 0	5 0	12½	42 0	3 6	10½	34 0	5 0	12	41 6	3 0	10	34 0	5 0	12	40 0	3 0	9½
Topgallant-mast and yard .	18 0	..	7½	31 6	2 0	7½	18 0	..	7	31 6	2 0	7	17 6	..	7	30 0	2 0	6½
Royal-mast and yard . .	12 0	..	4½	22 0	1 6	5	12 0	..	..	22 0	1 6	4½	12 0	Pole. 6 0	..	21 0	1 0	4½
Fore-mast and yard . . .	61 6	9 3	20	51 0	3 3	12¾	01 0	9 0	18½	51 0	3 0	12½	60 0	9 0	18½	50 0	3 0	12
Top-mast and yard . .	34 0	5 0	12½	41 6	3 6	10½	34 0	5 0	12	41 6	3 0	10	34 0	5 0	12	40 0	3 0	9½
Topgallant-mast and yard .	17 0	..	7½	29 6	2 0	7	18 0	..	7	31 6	2 0	7	17 6	..	7	30 0	2 0	0½
Royal-mast and yard . .	11 6	..	4½	20 0	1 3	4½	12 0	..	..	22 0	1 6	4½	12 0	Pole. 0 0	..	21 0	1 6	4½
Mizen-mast	61 0	8 9	14	..	..	..	60 0	8 0	13	..	..	..	60 0	8 0	13	..	..	..
Top-mast	39 6	..	8	..	..	..	48 0	..	8½	..	..	..	47 0	..	8	..	..	..
Pole	9 0	..	..	..	..	..	9 0	..	..	..	..	..	10 0	7 0	..	..	..	..
Bowsprit	37 0	..	22	..	..	..	35 0	..	19	..	..	..	34 0	..	19	..	..	..
Jib-boom	35 0	..	11	..	..	..	34 0	..	10½	..	..	..	34 0	..	10½	..	..	..
Flying-jibboom	30 0	..	8	..	..	..	38 0	..	6¾	..	..	..	38 0	..	6½	..	..	..
Mizen-boom	37 0	..	8	..	..	..	37 0	..	8	..	..	..	36 0	..	8	..	..	..
Gaff	31 0	5 0	8	..	..	..	29 0	4 0	7½	..	..	..	30 0	5 0	7½	..	..	..
Main-gaff	21 6	..	6½	..	..	..	19 0	..	8	..	..	..	18 0	..	8	..	..	..

DIMENSIONS OF MASTS AND YARDS IN THE MERCHANT SERVICE.

Names of the Masts and Yards.	Brig of 3[illegible] tons. Dimensions of Ship: Length 108 ft. 3 in.; Breadth, 27 ft. 6 in.						Barque of 1[illegible]0 tons. Dimensions of Ship: Length, 90 ft.; Breadth, 22 ft.						Brig of 280 tons. Dimensions of Ship: Length, 92 ft.; Breadth, 25 ft					
	Masts or Booms.			Yards.			Masts or Booms.			Yards.			Masts or Booms.			Yards.		
	Extreme Length.	Headed Length.	Diameter.	Extreme Length.	Arm.	Diameter.	Extreme Length.	Headed Length.	Diameter.	Extreme Length.	Arm.	Diameter.	Extreme Length.	Headed Length.	Diameter.	Extreme Length.	Arm.	Diameter.
	Ft. In.	Ft. In.	Ins.	Ft. In.	Ft. In.	Ins.	Ft. In.	Ft. In	Ins.	Ft. In.	Ft. In.	Ins.	Ft. In.	Ft. In.	Ins.	Ft. In.	Ft. In.	Ins.
Main-mast and yard	71 0	11 0	21½	55 0	3 0	14	54 0	8 0	16	42 0	2 6	10	60 0	9 0	18	50 0	3 0	11½
Top-mast and yard	40 0	6 6	12½	42 6	3 9	11	32 0	4 6	10	31 0	2 0	7¾	34 0	5 0	11½	40 0	3 6	9¼
Topgallant-mast and yard	19 6	..	7¾	28 0	2 0	7	16 0	1 6	6½	20 0	1 6	4½	17 0	..	7	30 0	2 0	6½
Royal-mast and yard	13 0	..	..	20 0	1 4	5	..	..	..	..	..	..	10 0	Pole. 8 0	4½	21 0	1 6	4½
Fore-mast and yard	66 0	11 0	21	51 6	3 0	13	51 0	8 0	16	42 0	2 6	10	58 0	9 0	18	50 0	3 6	11½
Top-mast and yard	38 0	6 0	12	39 0	3 6	10½	32 0	4 6	10	31 0	2 0	7¾	34 0	5 0	11½	40 0	3 6	9¼
Topgallant-mast and yard	18 6	..	7½	26 6	2 0	7	16 0	1 6	6½	20 0	1 6	4½	17 0	..	7	30 0	2 0	6½
Royal-mast and yard	12 0	..	..	13 0	1 3	5	..	..	..	..	..	..	10 0	Pole. 8 0	4½	21 0	1 6	4½
Mizen-mast	..	..	..	..	..	..	52 0	6 0	11	..	..	..	..	..	..	..	..	..
Top-mast	..	..	..	..	..	..	40 0	..	7	..	..	..	..	..	..	..	..	..
Pole	..	..	..	..	..	..	10 0	..	..	..	..	..	..	..	..	..	..	..
Bowsprit	42 0	..	22	..	..	..	32 0	..	16½	..	..	..	33 0	..	18	..	..	..
Jib-boom	38 0	..	11	..	..	..	33 0	..	9	..	..	..	34 0	..	10	..	..	..
Flying-jibboom	33 0	..	7	..	..	..	35 6	..	6	..	..	..	37 0	..	6¼	..	..	..
Mizen-boom or Main-boom	48 0	..	13	..	..	..	28 0	..	6	..	..	..	50 0	..	11	..	..	..
Gaff	..	..	..	..	..	..	25 0	3 0	6½	..	..	..	..	..	..	..	..	..
Main-gaff	34 0	4 0	8½	..	..	..	..	..	..	..	..	..	36 0	4 0	6½	..	..	..
Trysail-mast	..	..	..	..	..	..	..	..	..	..	..	..	37 0	..	..	..	..	..

DIMENSIONS OF MASTS AND YARDS IN THE MERCHANT SERVICE.

Names of the Masts and Yards.	Brig of 220 tons.						Brig of 250 tons.						Brig of 300 tons.					
	Masts or Booms			Yards			Masts or Booms			Yards			Masts or Booms			Yards		
	Extreme Length	Hoisted Length	Diameter	Extreme Length	Arm	Diameter	Extreme Length	Hoisted Length	Diameter	Extreme Length	Arm	Diameter	Extreme Length	Hoisted Length	Diameter	Extreme Length	Arm	Diameter
	Ft. In.	Ft. In.	Ins.	Ft. In.	Ft. In.	Ins.	Ft. In.	Ft. In.	Ins.	Ft. In.	Ft. In.	Ins.	Ft. In.	Ft. In.	Ins.	Ft. In.	Ft. In.	Ins.
Main-mast and yard	62 0	9 4	18	47 0	2 6	11	64 0	8 0	16	38 0	2 0	9	66 0	8 0	17½	40 0	2 9	10
Top-mast and yard	34 0	5 4	11	36 0	2 0	9	30 0	4 6	9½	30 0	1 6	7	31 0	4 0	10	32 0	2 0	8
Topgallant-mast and yard	18 0		6	26 0	1 6	6½	15 0		6	22 0	1 0	5	15 0		6	24 0	1 6	6
Royal-mast and yard	12 0		4½	16 0	1 0	4½	10 0		4				10 0		4	18 0	1 0	4
Fore-mast and yard	60 0	9 0	17¾	47 0	2 6	11	62 0	8 0	15½	38 0	2 0	9	64 0	8 0	16	40 0	2 9	10
Top-mast and yard	34 0	5 4	11	36 0	2 0	9	30 0	4 6	9½	30 0	1 6	7	31 0	4 6	10	32 0	2 0	8
Topgallant-mast and yard	18 0		6	29 0	2 6	6½	15 0		6	22 0	1 0	5	15 0		6	24 0	1 6	6
Royal-mast and yard	12 0		4½	21 0	1 6	4½	10 0		4				10 0		4	18 0	1 0	4
Bowsprit	36 0		18				34 0		17				34 0		17			
Jib-boom	33 0		8½				32 0		8½				32 0		8½			
Main-boom	44 0		12				37 0		10				38 0		10			
Gaff	26 0		9				29 0		7½				25 0		7½			
Housing of Fore-mast	16 6						13 3						13 6					
Main-mast	16 6						13 3						13 0					

DIMENSIONS OF MASTS AND YARDS IN THE MERCHANT SERVICE.

Names of the Masts and Yards.	Brig of 160 tons.						Brig of 10 keels.						Brigantine.					
	Masts or Booms.			Yards.			Masts or Booms.			Yards.			Masts or Booms.			Yards.		
	Extreme Length.	Headed Length.	Diameter.	Extreme Length.	Arm.	Diameter.	Extreme Length.	Headed Length.	Diameter.	Extreme Length.	Arm.	Diameter.	Extreme Length.	Headed Length.	Diameter.	Extreme Length.	Arm.	Diameter.
	Ft. In.	Ft. In.	Ins.	Ft. In.	Ft. In.	Ins.	Ft. In.	Ft. In.	Ins.	Ft. In.	Ft. In.	Ins.	Ft. In.	Ft. In.	Ins.	Ft. In.	Ft. In.	Ins.
Main-mast and yard	48 0	7 0	14½	39 0	3 0	10	43 0	8 0	14½	37 0	2 9	9½	53 0	6 6	13	..	..	..
Top-mast and yard	28 0	4 0	9	31 0	3 0	8	26 6	4 6	8½	29 0	2 8	7½	34 0	..	7½	..	..	..
Topgallant-mast and yard	14 0	..	4½	24 0	1 6	6	13 6	..	4½	21 0	1 9	5	..	..	..	..	..	..
Royal-mast and yard	9 6	..	4	18 0	1 0	4	8 6	Pole. 5 0	4	15 0	1 3	4	..	..	..	..	..	..
Fore-mast and yard	46 0	7 0	14½	39 0	3 0	10	46 0	8 0	14½	37 0	2 9	9½	49 0	6 0	15	40 0	2 6	10
Top-mast and yard	28 0	4 0	9	31 0	3 0	8	26 6	4 6	8½	29 0	2 8	7½	24 6	3 0	8	31 0	2 0	8
Topgallant-mast and yard	14 0	..	4½	24 0	1 6	6	13 6	..	4½	21 0	1 9	5	11 0	..	4	22 0	1 6	5½
Royal-mast and yard	9 6	..	4	18 0	1 0	4	8 6	Pole. 5 0	4	15 0	1 3	4	7 0	..	3½	16 6	1 0	4
Bowsprit	28 0	..	14	..	..	..	29 0	..	14½	..	..	..	26 0	..	13½	..	..	..
Jib-boom	26 0	..	6¾	..	..	..	27 9	..	7	..	..	..	26 0	..	6¾	..	..	..
Main-boom	33 0	..	9½	..	..	..	34 0	..	8½	..	..	..	37 6	..	9½	..	..	..
Gaff	27 0	..	7½	..	..	..	28 0	..	7½	..	..	..	30 0	End. 4 0	8	..	..	..
Fore-boom	..	..	..	..	..	..	..	..	..	..	..	..	22 0	..	6	..	..	..
Gaff	..	..	..	..	..	..	..	..	..	..	..	..	20 0	..	5½	..	..	..
Housing of Masts	12 0	..	..	..	..	..	11 0	..	..	..	..	..	10 8	..	..	..	..	..

DIMENSIONS OF MASTS, YARDS, ETC., OF A FAST-SAILING CLIPPER SCHOONER.

	Ft. In.		Ft. In.		Ins.
Main-mast from deck to top of cheeks	47 10	Head .	8 3	Diameter	15½
Fore-mast do. do. .	45 0	,, .	7 10	,,	15
Fore-top-mast, hoist	21 0 •	.. ..		,,	7
Fore-topgallant-mast . . .	12 0	Royal .	8 6	,,	5
Main-top-mast, hoist . . .	25 0	Pole .	6 0	,,	7½
Bowsprit outside	20 0	.. ..		,,	15
Jib-boom outside of cap . .	16 0	.. ..		,,	6
Flying-jibboom	10 6	.. ..		,,	3½
Fore-yard	55 0	Arm	2 10	,,	13
Fore-topsail-yard	41 0	,, .	2 3	,,	10
Fore-topgallant-yard . . .	29 6	,, .	1 6	,	7
Main-boom	59 0	.. ..		,,	13
Gaff for sail	29 0	Pole .	4 0	,,	7
Fore-gaff	23 3	.. ..			6
Gaff-topsail-yard for sail . .	7 0	.. ..		,,	3
Lower-masts, house-each . .	13 6	.. ..		..	..
Distance from fore-stay to centre of fore-mast	29 6	.. ..		..	..
Distance from centre of fore-mast to main-mast	34 0	.. ..		..	..
Distance from centre of main-mast to taffrail	46 6	.. ..		..	..
Breadth of rigging to rigging at fore-mast	21 0	.. ..		..	..
Breadth of rigging to rigging at main-mast	21 0	.. ..		..	..
Height of rail	3 6	.. ..		..	..
Rise of deck	1 0	.. ..		..	..
Rake of the fore-mast to the foot .	0 1½	.. ..		..	..
Rake of the main-mast to the foot .	0 2	.. ..		..	..
Steave of bowsprit to the foot . .	0 8½	.. ..		..	..

DIMENSIONS OF MASTS AND YARDS OF SCHOONERS.

NAMES OF THE MASTS AND YARDS	Schooner of 143 tons.						Schooner of 91 tons.					
	Masts or Booms.			Yards.			Masts or Booms.			Yards.		
	Extreme Length	Headed Length	Diameter	Extreme Length	Arm	Diameter	Extreme Length	Headed Length	Diameter	Extreme Length	Arm	Diameter
	Ft. In.	Ft. In.	Ins.	Ft. In.	Ft. In.	Ins.	Ft. In.	Ft. In.	Ins.	Ft. In.	Ft. In.	Ins.
Main-mast	58 0	7 0	13	:	:	:	54 0	8 0	12½	:	:	:
Top-mast and pole	33 0	8 0	7½	:	:	:	28 0	6 0	7½	:	:	:
Fore-mast and yard	57 0	7 0	13	41 0	2 0	8	40 0	7 2	12	36 0	1 6	7½
Top-mast and yard	18 0	:	7½	33 0	1 10	7¼	24 0	4 0	7	23 0	1 4	6
Topgallant-mast and yard	11 0	:	:	22 0	1 6	5¼	12 0	:	4½	18 0	1 0	4½
Bowsprit	27 0	:	12½	:	:	:	27 0	:	12	:	:	:
Jib-boom	32 0	:	7½	:	:	:	26 0	:	7	:	:	:
Pole	6 0	:	:	:	:	:	:	:	:	:	:	:
Main-boom	40 0	:	10	:	:	:	36 0	:	9	:	:	:
Gaff	28 0	:	7½	:	:	:	28 0	:	7½	:	:	:
Fore-boom	24 0	:	6½	:	:	:	24 0	:	6½	:	:	:
Gaff	21 0	:	6	:	:	:	22 0	:	6¼	:	:	:

DIMENSIONS OF MASTS AND YARDS OF SCHOONERS.

NAMES OF THE MASTS AND YARDS.	Schooner of 187 tons. Dimensions of Vessel: Length, 81 ft.; Breadth, 22 ft.						Schooner of 168 tons. Dimensions of Vessel: Length, 79 ft.; Breadth, 21 ft.						Schooner of 100 tons. Dimensions of Vessel: Length, 75 ft.; Breadth, 21 ft.					
	Masts or Booms.			Yards.			Masts or Booms.			Yards.			Masts or Booms.			Yards.		
	Extreme Length.	Headed Length.	Diameter.	Length.	Arm.	Diameter.	Extreme Length.	Headed Length.	Diameter.	Extreme Length.	Arm.	Diameter.	Extreme Length.	Headed Length.	Diameter.	Extreme Length.	Arm.	Diameter.
	Ft. In.	Ft. In.	Ins.	Ft. In.	Ft. In.	Ins.	Ft. In.	Ft. In.	Ins.	Ft. In.	Ft. In.	Ins.	Ft. In.	Ft. In.	Ins.	Ft. In.	Ft. In.	Ins.
Main-mast	64 6		15				58 0	7 0	13				39 0	7 0	13			
From deck to cheeks		9 0											28 0	6 0	9			
Top-mast and pole	28 0	6 0	7	54 0	2 8	9	38 0	8 0	8½	38 0	2 6	8	47 0	8 0	13	35 0	1 0	7
Fore-mast and sq.-sail yard	62 0	9 0	15	34 0	2 0	7	49 0	6 6	12	30 0	1 9	7¼	16 0	6 0	6¼	31 0	1 0	7
Top-mast and yard	20 0		7½	26 0	1 8	6	26 0	8 9	7¼	21 0	1 6	5	9 0			24 0	0 9	5¼
Topgallant-mast and yard	10 0	3 0					13 0	8 0					23 6	3 0	15			
Bowsprit	29 0		16				28 0		13				22 0		7			
Jib-boom	30 0		8½				28 0		7½									
Pole													38 0		9			
Main-boom	40 0		10½				39 0		10				27 0		7			
Gaff	24 0		7				27 0		7½				25 0		7			
Fore-boom													25 6		5	10 0		2
Gaff															6¼			
Fore-topmast, studding-sail, boom, and yard													12 0		3	7 0		2
Fore-topgallant-mast, studding-sail, boom, and yard													23 0		6¼	9 0		2¼
Swinging-boom and yard																		

DIMENSIONS OF MASTS AND YARDS OF SCHOONERS.

Names of the Masts and Yards	Schooner of 5 keels. Masts or Booms. Extreme Length (Ft. In.)	Headed Length (Ft. In.)	Diameter (Ins.)	Yards. Extreme Length (Ft. In.)	Arm (Ft. In.)	Diameter (Ins.)	Schooner of 50 tons. Masts or Booms. Extreme Length (Ft. In.)	Headed Length (Ft. In.)	Diameter (Ins.)	Yards. Extreme Length (Ft. In.)	Arm (Ft. In.)	Diameter (Ins.)
Main-mast	40 0	6 6	13	...	...	...	47 0	8 0	11	...	...	...
Top-mast and pole	29 0	5 6	7½	...	...	...	26 0	6 0	7	...	1 0	6½
Fore-mast and yard	48 0	6 6	12	35 0	1	7½	46 0	10 0	12	33 0	1 2	5¾
Top-mast and yard	15 0	...	7½	27 0	1 4	0	26 0	4 0	7½	27 0	0 9	4½
Topgallant-mast and yard	7 8	...	...	19 0	1 0	4½	11 0	...	4½	19 0	...	...
Pole	5 6	...	...	...	...	...	...	...	...	...	...	...
Bowsprit	24 0	Pole.	11	...	...	...	25 0	...	11¼	25 0	...	...
Jib-boom	25 0	3 0	7	...	...	...	23 0	...	6½	23 0	...	...
Main-boom	33 0	...	7½	...	...	...	31 0	...	7½	31 0	...	...
Gaff	24 6	...	8½	...	...	...	21 0	...	7	21 0	...	...
Fore-boom	20 0	...	6½	...	...	...	23 0	...	6½	23 0	...	...
Gaff	18 0	...	6	...	...	...	20 0	...	0	20 0	...	...

DIMENSIONS OF MASTS AND YARDS OF AMERICAN VESSELS.

Names of the Masts and Yards.	Barks. Masts or Booms. Extreme Length.	Barks. Masts or Booms. Headed Length.	Barks. Masts or Booms. Diameter.	Barks. Yards. Extreme Length.	Barks. Yards. Arm.	Barks. Yards. Diameter.	Brigantines. Masts or Booms. Extreme Length.	Brigantines. Masts or Booms. Headed Length.	Brigantines. Masts or Booms. Diameter.	Brigantines. Yards. Extreme Length.	Brigantines. Yards. Arm.	Brigantines. Yards. Diameter.	*Schooners. Masts or Booms. Extreme Length.	*Schooners. Masts or Booms. Headed Length.	*Schooners. Masts or Booms. Diameter.	*Schooners. Yards. Extreme Length.	*Schooners. Yards. Arm.	*Schooners. Yards. Diameter.
	Ft. In.	Ft. In.	Ins.	Ft. In.	Ft. In.	Ins.	Ft. In.	Ft. In.	Ins.	Ft. In.	Ft. In.	Ins.	Ft. In.	Ft. In.	Ins.	Ft. In.	Ft. In.	Ins.
Main-mast and yard	72 2	12 2	22½	59 7	3 0	14	76 0	8 0	20	35 7	2 10	8	78 0	8 0	20¼	..	..	..
Top-mast and yard	40 7	6 10	12½	44 8	3 9	10¾	21 0	..	6¼	24 7	1 5	5½	26 2	..	7½	..	..	..
Topgallant-mast and yard	20 3	..	7	28 5	1 5	6½	14 0	..	5¼	16 3	0 10	3½	13 1	..	5	..	..	..
Royal-mast and yard	13 4	..	..	13 0	0 10	3¾	0 7	..	..	..	..	..	..	..	..	..	..	..
Flag-pole	5 5	..	Truck 2½	..	..	..	5 5	..	..	..	..	..	6 0	..	Truck 2½	..	..	..
Fore-mast and yard	04 9	11 3	22	59 7	3 0	14	65 0	9 2	18¾	45 0	2 10	10	73 9	8 0	21¼	50 0	2 7	11½
Top-mast and yard	40 7	6 10	12½	44 8	3 9	10¾	32 7	5 7	10½	33 7	3 0	7¾	26 2	..	7½	33 5	2 7	7½
Topgallant-mast and yard	20 3	..	7	28 5	1 5	6½	18 3	..	6¾	22 0	1 7	4½	13 1	..	5	22 2	1 0	4½
Royal-mast and yard	13 4	..	..	13 0	0 10	3¾	12 0	..	..	14 10	0 10	3	..	..	..	..	..	..
Flag-pole	5 5	..	Truck 2½	..	..	..	5 0	..	..	..	..	..	6 0	..	Truck 2½	..	..	..
Bowsprit	42 0	..	21	..	..	..	17 7	Out	b'rd	..	..	..	29 0	..	18	..	..	..
Jib-boom	32 0	..	9½	..	..	..	14 0	Out	b'rd	..	..	..	37 0	..	11	..	..	..
Flying-jibboom	34 6	..	7	..	..	..	12 0	Out	b'rd	..	..	..	..	..	..	..	..	..
Pole	5 5	..	..	..	..	..	8 5	..	..	..	..	..	..	..	..	..	..	..
Spanker-boom	59 7	..	13	..	..	..	50 0	..	11	..	..	..	50 0	..	11	..	..	..
Main-gaff	39 9	Pole. 5 0	10	..	..	..	25 0	Pole. 5 0	8½	..	..	..	25 0	Pole. 5 0	8½	..	..	..
Fore-gaff	25 0	..	6¼	..	..	..	24 0	..	8	..	..	..	25 0	..	8	..	..	..
Square-sail-boom and yard	..	..	..	..	..	..	..	..	..	..	..	..	45 0	..	9¾	22 9	..	5
Main-top-mast stud.-sail-boom	29 0	..	0	12 0	..	2¾	..	..	..	..	..	..	..	..	..	..	..	..
Topgallant-mast stud.-sail-boom and yard	22 3	..	4½	12 0	..	2¾	..	..	..	..	..	..	..	..	..	..	..	..
Lower-swinging-boom and yard	31 7	..	6½	15 3	..	3	22 0	..	4½	11 0	..	2¾	22 0	..	4¾	11 0	..	2¾
Fore-top-mast stud.-sail-boom and yard	29 0	..	6	12 0	..	2¾	22 7	..	4¾	13 10	..	3	25 0	..	5¼	14 2	..	3
Topgallant-mast stud.-sail-boom and yard	22 3	..	4¾	12 0	..	2¼	16 10	..	3¾	9 7	..	2	16 9	..	3¾	10 0	..	2¼

DIMENSIONS OF MASTS AND YARDS OF STEAM-VESSELS.

Names of the Masts and Yards	Iron Steamer of 700 tons. Dimensions of Ship; Length, 170 ft.; Breadth, 29 ft.						Steam-vessel of 350 tons. Dimensions of Ship; Length, 140 ft.; Breadth, 22 ft. 6 in.						Iron Steamer of 300 tons. Dimensions of Ship; Length, 140 ft.; Breadth, 20 ft. 6 in.					
	Masts or Booms			Yards			Masts or Booms			Yards			Masts or Booms			Yards		
	Extreme Length.	Headed Length.	Diameter.	Extreme Length.	Arm.	Diameter.	Extreme Length.	Headed Length.	Diameter.	Extreme Length.	Arm.	Diameter.	Extreme Length.	Headed Length.	Diameter.	Extreme Length.	Arm.	Diameter.
	Ft. In.	Ft. In.	Ins.	Ft. In.	Ft. In.	Ins.	Ft. In.	Ft. In.	Ins.	Ft. In.	Ft. In.	Ins.	Ft. In.	Ft. In.	Ins.	Ft. In.	Ft. In.	Ins.
Main-mast and yard	68 0	8 0	15	52 0	3 0	10½	56 0	9 0	14½	41 0	1 10	11½	55 6	7 8	13½	..	..	..
Top-mast and yard	25 0	4 0	10	36 0	2 0	8½	40 6	..	7½	31 0	3 0	5¼	44 0	..	7	..	..	..
Topgallant-mast and yard	26 0	8 0	..	25 0	1 6	6	6 0	..	..	18 6	0 10	4	9 0	..	..	..	..	..
Fore-mast and yard	70 0	8 0	15	52 0	3 0	10½	53 9	7 8	15	41 0	1 10	11½	53 6	7 8	13½	42 0	2 3	8
Top-mast and yard	28 0	4 0	10	36 0	2 0	8½	37 0	..	8¼	31 0	3 0	5¼	44 6	..	7½	31 0	4 0	6
Topgallant-mast and yard	27 0	8 0	..	25 0	1 6	6	10 6	1 10	..	18 6	0 10	4	13 0	7 0	..	21 6	1 3	4½
Bowsprit	30 0	..	15	..	..	..	30 6	12 0	13	..	..	..	27 0	..	11	..	..	..
Jib-boom	13 0	..	8	..	..	..	27 6	..	6¼	..	..	..	30 0	..	7½	..	..	..

DIMENSIONS OF MASTS AND YARDS OF STEAM-VESSELS.

NAMES OF THE MASTS AND YARDS	Steam Ship of 1250 tons. Dimensions of Ship: Length, 200 ft.; Breadth, 43 ft.						Steam Ship of 1915 tons. Dimensions of Ship: Length, 195 ft.; Breadth, 38 ft.						Steam Ship of 863 tons. Dimensions of Ship: Length, 180 ft.; Breadth, 33 ft.					
	Masts or Booms			Yards			Masts or Booms			Yards			Masts or Booms			Yards		
	Extreme Length	Headed Length	Diameter	Extreme Length	Arm	Diameter	Extreme Length	Headed Length	Diameter	Extreme Length	Arm	Diameter	Extreme Length	Headed Length	Diameter	Extreme Length	Arm	Diameter
	Ft. In.	Ft. In.	Ins.	Ft. In.	Ft. In.	Ins.	Ft. In.	Ft. In.	Ins.	Ft. In.	Ft. In.	Ins.	Ft. In.	Ft. In.	Ins.	Ft. In.	Ft. In.	Ins.
Main-mast and yard	106 0	14 0	20½	86 0	3 6	19	90 0	12 0	22	50 0	2 1	9½	74 0	12 0	10	56 0	2 4	12¼
Top-mast and yard	67 0	..	13½	66 0	8 0	13	57 0	..	11	38 6	5 0	8½	40 6	..	8½	46 0	3 10	7½
Topgallant-mast and yard	12 0	..	..	39 0	1 8	8¼	10 0	..	..	29 6	1 3	6¼	8 0	..	..	21 0	1 0	5½
Fore-mast and yard	100 6	14 0	20	88 0	8 6	19	80 0	12 6	22½	64 0	2 8	14½	68 6	10 0	17½	..	..	..
Top-mast and yard	63 6	..	15	66 0	8 9	13	51 0	..	12	50 0	6 8	9½	40 0	..	10	..	..	..
Topgallant-mast and yard	19 6	8 0	..	39 0	1 8	8½	16 6	4 6	..	29 6	1 8	6½	13 0	2 0	..	..	..	..
Mizen-mast	57 6	10 0	17	..	..	..	..	..	..	..	..	..	..	..	..	..	..	..
Top-mast	52 0	..	10	..	..	..	..	..	..	..	..	..	..	..	..	..	..	..
Topgallant-pole	8 0	..	..	..	..	..	..	..	..	..	..	..	..	..	..	..	..	..
Bowsprit	48 8	..	26	..	..	..	38 6	..	19	..	..	..	32 6	..	14	..	..	..
Jib-boom	46 6	..	13	..	..	..	35 0	..	19	..	..	..	31 0	..	7½	..	..	..

DIMENSIONS OF MASTS AND YARDS OF AN AMERICAN STEAM-SHIP OF 630 TONS.

	Ft. In.	Ft. In.	Ft. In.	In.
Main-mast	74 0	Head . 12 0	.. ..	Diameter 24½
Fore-mast	70 0	,, . 12 0	.. ..	,, 25
Mizen-mast	54 6	,, . 10 0	.. ..	,, 13½
Fore and main top-mast	42 0	,, . 7 6	.. ..	Cap . . 14½
Topgallant-mast	22 0	Hoist . 14 6	Royal . 9 0	Pole-cap. 8½
Mizen-top-mast	33 0	Head . 5 6	.. ..	Cap . . 11
Top-gallant-mast	18 0	Hoist . 12 0	Royal . 7 0	Pole . 6¾
Main and fore-yard	68 0	.. ..	Arms . 4 0	Sling . . 15½
Topsail-yard	54 0	.. ..	,, . 4 7	,, . 13
Topgallant-yard	37 0	.. ..	,, . 2 7	,, . 7¾
Royal-yard	25 7	.. ..	,, . 1 8	,, . 5½
Mizen-yard	54 0	.. ..	,, . 4 7	,, . 13
Topsail-yard	39 0	.. ..	,, . 3 7	,, . 9
Topgallant-yard	26 7	.. ..	,, . 1 11	,, . 6½
Royal-yard	19 0	.. ..	,, . 1 0	,, . 4
Bowsprit, outboard	25 7	.. ..	.. ..	Bod . . 25
Jib-boom, ditto	24 0	Inboard 20 0	Head . 3 0	Cap . . 13½
Flying-jib-boom, ditto	19 0	.. ..	,, . 5 0	,, . 8½
Spanker-boom	47 7	.. ..	,, . 2 0	Diameter 9
Gaff	30 0	.. ..	,, . 8 0	,, 7
Swinging-booms	45 7	.. ..	.. ..	,, 8
Top-mast stud.-sail-booms	35 0	.. ..	.. ..	,, 7
Topgallant ditto	28 0	.. ..	.. ..	,, 5½
Royal ditto	19 0	.. ..	.. ..	,, 3¾
Lower ditto	17 0	.. ..	.. ..	,, 5
Top-mast ditto	21 0	.. ..	.. ..	,, 5
Topgallant ditto	16 0	.. ..	.. ..	,, 4
Royal ditto	10 0	.. ..	.. ..	,, 3
Mizen trysail-mast	38 0	.. ..	.. ..	,, 7½
Housing of Fore-mast	20 5			
Main-mast	20 8			
Mizen-mast	7 1			
Length between perpendiculars	156 0			
Spar-deck	165 2			
Keel	142 10			
Extreme breadth of beam	30 0			

DIMENSIONS OF MASTS AND YARDS OF STEAM-VESSELS.

Name of the Masts and Yards	Iron Steamer of 230 tons. Length, 143 ft.; Breadth, 21 ft.						Steamer of 250 tons. Length, 125 ft. 6 in.; Breadth, 22 ft.						Iron Steamer of 180 tons. Length, ... ft.; Breadth, 20 ft.					
	Masts or Booms			Yards			Masts or Booms			Yards			Masts or Booms			Yards		
	Extreme Length Ft. In.	Headed Length Ft. In.	Diameter In.	Extreme Length Ft. In.	Arm Ft. In.	Diameter In.	Extreme Length Ft. In.	Headed Length Ft. In.	Diameter In.	Extreme Length Ft. In.	Arm Ft. In.	Diameter In.	Extreme Length Ft. In.	Headed Length Ft. In.	Diameter In.	Extreme Length Ft. In.	Arm Ft. In.	Diameter In.
Main-mast	33 0		13				65 0		14				33 0		13			
From deck to hounds	26 6	7 0					30 6	8 0					21 9	6 0	6½			
Top-mast to stops			9				32 10		7									
Pole	5 0		6				5 0						4 6					
Fore-mast and yard	34 0		18	40 0	2 0	11	54 6		14	40 0	1 10	11	33 0		13	38 0	2 0	10
From deck to hounds	29 0	7 0	5½	31 6	2 0	6½	35 6	8 0	7	30 8	3 6	6½	18 0	6 0	5	29 0	1 6	6
Top-mast to stops and yard	0 0		5	23 0	1 0	5	24 0		5	18 0	0 10	5	9 0		4½	21 0	1 0	5
Topgallant-mast and yard							10 0											
Royal pole	1 8						1						1 6					
Mizen-mast from deck to hounds	32 0		10															
Pole	14 0																	
Bowsprit, outboard	10 6		13				17 0		13				12 0		12			
Jib-boom, ditto	10 0		4				16 0		6				18 0		7			
Main-boom	30 0		7½				46 0		10				35 0		9			
Fore-boom	33 0		8										26 0		6½			
Mizen-boom	24 0		6															
Main-gaff	29 0		6				23 0		6				26 6		6½			
Fore-gaff	29 0		6				23 0		6				21 3		6			
Mizen-gaff	19 0		5															
Distance from stem to fore-mast	35 0						20 0											
From centre of fore-mast to main-mast	66 6						57 0											
From centre of main-mast to mizen-mast	51 6						42 6											
From centre of mizen-mast to taffrail	12 6																	

DIMENSIONS OF MASTS AND YARDS OF YACHTS.

Names of the Masts and Yards	Schooner Yacht of 177 tons. Dimensions of Vessel: Length, 90 ft.; Breadth, 22 ft.						Schooner Yacht of 170 tons. Dimensions of Vessel: Length, 94 ft.; Breadth, 20 ft. 6 in.						Schooner Yacht of 75 tons. Dimensions of Vessel: Length, 77 ft.; Breadth, 18 ft. 8 in.					
	Masts or Booms.			Yards.			Masts or Booms.			Yards.			Masts or Booms.			Yards.		
	Extreme Length. Ft. In.	Headed Length. Ft. In.	Diameter. In.	Extreme Length. Ft. In.	Arm. Ft. In.	Diameter. In.	Extreme Length. Ft. In.	Headed Length. Ft. In.	Diameter. In.	Extreme Length. Ft. In.	Arm. Ft. In.	Diameter. In.	Extreme Length. Ft. In.	Headed Length. Ft. In.	Diameter. In.	Extreme Length. Ft. In.	Arm. Ft. In.	Diameter. In.
Main-mast	73 0	7 4	13				64 0	8 6	24				52 0	6 9	17			
From deck to hounds .	24 0		7				50 0		9				29 0		9			
Top-mast	8 7												9 0					
Top-mast-pole . . .	73 0	7 4	19	55 0		9	62 3	8 0	23				60 0	6 6	17	43 0	2 0	8
Fore-mast and yard .	18 6		7½	29 2	2 3	7	52 0		9	17 3			18 0		7½	30 0	2 3	7
From deck to hounds .	8 0	31 0	6	23 0	4 0	6							10 0		6	20 0	1 0	6
Top-mast to stops and yard	4 0		17				17 3		13				8 0		17			
Topgallant-mast and yard	12 0		10				23 0		8				15 0		6			
Topgallant-pole . .	30 0		11				53 0		12				13 6		5			
Bowsprit, outboard .	50 5												44 0		10			
Jib-boom, ditto . .																		
Flying-jib-boom, ditto .																		
Main-boom . . .								Pole. 1 6							Pole. 2 0				
Main-gaff	24 4		8				28 0		8				27 6		8			
Fore-gaff	23 0		9				26 9		8				26 0		8			
Gaff-topsail-gaff . .													7 6		3			

* The celebrated American Yacht.

DIMENSIONS OF MASTS, ETC., OF YACHTS.

Names of the Masts and Yards.	Sloop Yacht of 125 tons. Dimensions of Vessel: Length, 63 ft. 9 in.; Breadth, 20 ft. 6 in.						Cutter of 108 tons. Dimensions: Length, 60 ft.; Breadth, 20 ft. 6 in.			Cutter of 64 tons. Dimensions: Length, 45 ft.; Breadth, 14 ft. 6 in.			Cutter of 40 tons. Dimensions of Vessel: Length, 40 ft.; Breadth, 14 ft.					
	Masts or Booms.			Yards.			Masts or Booms.			Masts or Booms.			Masts or Booms.			Yards.		
	Extreme Length (Ft. In.)	Headed Length (Ft. In.)	Diameter (Ins.)	Extreme Length (Ft. In.)	Arm (Ft. In.)	Diameter (Ins.)	Extreme Length (Ft. In.)	Headed Length (Ft. In.)	Diameter (Ins.)	Extreme Length (Ft. In.)	Headed Length (Ft. In.)	Diameter (Ins.)	Extreme Length (Ft. In.)	Headed Length (Ft. In.)	Diameter (Ins.)	Extreme Length (Ft. In.)	Arm (Ft. In.)	Diameter (Ins.)
Mast	74 2	13 0	20				68 0	11 0	17½	63 0	11 6	17½	47 6	6 6	12			
Top-mast to stops	35 0		7½				40 0		8½	37 6		8	25 0		6½			
Top-mast-pole	4 0									8 0			4 0					
Bowsprit	33 8		14½				48 0		15	46 6		12	30 0		10½			
Main-boom	52 6		11				60 0		15	52 6		12	40 0		9½			
Main-gaff	33 0		8				48 0		12	37 0		9	25 6		7			
Square-sail-yard				38 0	1 0	7										32 0	1 0	6
Top-sail-yard				28 0	1 6	5½												
Winter-boom							47 0		12									
Winter-gaff							30 0		9									
Gaff-top-sail-yard													16 6					5¾
Outside of stem to centre of mast																15 0		
Centre of mast to the taffrail													32 0					

PROPORTIONS GIVEN FOR MASTS AND GEAR OF BOATS DIFFERENTLY RIGGED.

Sliding Gunter Sails.

Length of boat, 32 ft.; breadth, 8 ft. 6 in. or 8·5 ft.

Species of Masts and Gear.		Known quantities.			Proportions.	Length.	Diameter.
Main-mast	Is equal the	Breadth	8·5	Multiplied by	2·2	= 18·7 feet	6¾ inches.
Fore-mast		Main-mast	18·7		·93	= 17·5 ,,	6¼ ,,
Mizen-mast		Ditto	18·7		·53	= 10·0 ,,	3¾ ,,
Main-slide		Ditto	18·7		1·0	= 18·7 ,,	4½ ,,
Fore-slide		Fore-mast	17·5		1·	= 17·5 ,,	4¼ ,,
Mizen-slide		Mizen-mast	10·0		1·	= 10·0 ,,	2½ ,,
Bowsprit		Length	32·0		·25	= 8·0 ,,	5 ,,
Outrigger		Ditto	32·0		·34	= 11·0 ,,	4 ,,
Main-mast from the middle		Ditto	32·0		·053	= 1·7 before.	
Fore-mast ditto		Ditto	32·0		·328	= 10·5 ,,	

Main-mast to rake, in a foot, 1¼ inches.
Fore-mast ditto ditto 1 inch.
Mizen-mast ditto ditto as the transom.

Lateen or Setter Sails.

Length of boat, 32 ft.; breadth, 8·5 ft.

Species of Masts and Gear.		Known quantities.			Proportions.	Length.	Diameter.
Main-mast	Is equal the	Breadth	8·5	Multiplied by	2·15	= 18·3 feet	6¾ inches.
Fore-mast		Main-mast	18·3		·8	= 14·6 ,,	5½ ,,
Mizen-mast		Ditto	18·3		·43	= 8·0 ,,	4 ,,
Main-yard		Length	32·0		·83	= 26·5 ,,	6 ,,
Fore-yard		Main-yard	26·5		·95	= 25·0 ,,	5¾ ,,
Mizen-yard		Ditto	26·5		·56	= 15·0 ,,	3½ ,,
Outrigger		Length	32·0		·38	= 12·2 ,,	4½ ,,
Main-mast from the middle		Ditto	32·0		·037	= 1·2 abaft.	
Fore-mast ditto		Ditto	32·0		·312	= 10·0 before.	

PROPORTIONS GIVEN FOR MASTS AND GEAR OF BOATS DIFFERENTLY RIGGED.

CUTTER WITH THREE LUGSAILS, SQUARE AT THE HEADS.

Length of boat, 26 ft.; breadth, 6·5 ft.

Species of Masts and Gear.	Known quantities.		Proportions.	Length.	Diameter.
Main-mast	Breadth	6·5	·4	15·6 feet	4 inches
Fore-mast	Main-mast	15·6	·92	14·3 ,,	3⅛ ,,
Mizen-mast	Ditto	15·8	·6	9·3 ,,	2¼ ,,
Main-yard	Length	26·0	·5	13·0 ,,	3¼ ,,
Fore-yard	Main-yard	13·0	·9	11·7 ,,	2⅛ ,,
Mizen-yard	Ditto	13·0	·63	8·2 ,,	2¼ ,,
Main-mast from the middle	Length	20·0	·04	·8 before.	
ditto	Ditto	20·0	·037	7·4 ,,	
Outrigger	Ditto	26·0	·4	10·4 feet	4 ,,

Is equal the · Multiplied by

CUTTER WITH THREE LUGSAILS, NARROW AT THE HEADS.

Length of boat, 28 ft.; breadth, 7 ft.

Species of Masts and Gear.	Known quantities.		Proportions.	Length.	Diameter.
Main-mast	Breadth	7·0	·7	18·0 feet	4½ inches
Fore-mast	Main-mast	18·9	·9	17·0 ,,	4 ,,
Mizen-mast	Ditto	18·0	·6	11·3 ,,	2¾ ,,
Main-yard	Length	28·0	·38	10·6 ,,	2⅞ ,,
Fore-yard	Main-yard	10·6	·80	9·1 ,,	2¼ ,,
Mizen-yard	Ditto	10·6	·65	5·8 ,,	1½ ,,
Outrigger	Length	28·0	·34	9·5 ,,	4 ,,
Main-mast from the middle	Ditto	28·0	·015	·4 abaft.	
ditto	Ditto	28·0	·281	7·8 before.	

Is equal the · Multiplied by

Main-mast to rake, in a foot, 1 inch.
Fore-mast ditto ditto ½ inch.
Mizen-mast ditto ditto as the transom.

PROPORTIONS GIVEN FOR MASTS AND GEAR OF BOATS DIFFERENTLY RIGGED.

Gig, with Fore and Main Lugsails.

Length of boat, 26 ft.; breadth, 6 ft.

Species of Masts and Gear.	Known quantities.	Proportions.	Length.	Diameter.
Main-mast	6·0	2·7	= 16·2 feet	4 inches.
Fore-mast	16·2	·9	= 14·5 "	3½ "
Main-yard	26·0	·38	= 10·6 "	2¾ "
Fore-yard	10·6	·86	= 9·1 "	2¼ "
Main-mast from the middle	26·0	·015	= ·4 abaft.	
Fore-mast ditto	26·0	·291	= 7·8 before	

Breadth is the breadth — Main-mast, Length.
Main-yard is the length — Length, Ditto.

Multiplied by

Main-mast to rake, in a foot, 1 inch.
Fore-mast ditto ditto ½ inch.

Gig, with One Lugsail.

Length of boat, 20 ft.; breadth, 5 ft. 6 in. or 5·5 ft.

Mast is equal the breadth . . . 5·5 multiplied by 2·7 = 14·85 feet long . . . Diameter 3½ inches

Yard is equal the length . . . 20·0 " ·5 = 10· feet long . " 2½ "

Mast from middle, length . . . 20·0 " ·4 = 8· feet before.

END OF RUDIMENTARY MASTING OF SHIPS.

To commence with *Rigging*, it is necessary that the young student should practise the following, while preparing for a nautical life.

TO MAKE AN OVERHANDED KNOT.

To make an overhanded knot, pass the end of the rope over the standing part and through the bight, as the annexed sketch.

FIGURE OF EIGHT KNOTS.

Take the end of the rope round the standing part, under its own part and through the lower bight, and the knot is made.

SQUARE OR REEF KNOT.

First make an overhanded knot, supposing it be round a yard; then bring the end being next to you over the left hand and through the bight; haul both ends taut, and it is made as per sketch.

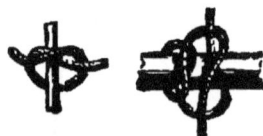

TO MAKE A BOWLINE KNOT.

Take the end of the rope in your right hand, and the standing part in the left; lay the end over the standing part, then with your left hand turn the bight of the standing part over the end part; then lead the end through the standing part above, and stick it down through the cuckold's neck formed on the standing part, and it will appear as the sketch.

TO MAKE TWO HALF-HITCHES.

Pass the end of the rope round the standing part, and bring it up through the bight—this is one half-hitch; two of these, one above the other, constitute two half-hitches, as the annexed figure.

A TIMBER-HITCH.

Take the end of the rope round a spar; pass it under and over the standing part, then pass several turns round its own part and it is done. The bight serves as a sling for bales, drawing of timber, &c.

A ROLLING-HITCH.

With the end of a rope take a half-hitch around the standing part; then take another through the same bight, jamming it above the first hitch and the upper part of the bight, then haul it taut, and lay the end above the hitch around the standing part, and stop the end back with a yarn.

A BLACKWALL-HITCH.

To make a Blackwall over a hook, you form a bight or a "kink" with the rope, having it underneath and the hook on the top; stick the hook through the bight, keeping the bight well up on the back of the hook (as shown in the figure), until the tackle is set taut. This is better learned by practice than it can be described.

BOWLINE UPON THE BIGHT OF A ROPE.

Take the bight of the rope in one hand and the standing part in the other; throw a cuckold's neck or a kink over the bight with the standing parts, the same as for the single knot. Take the bight round the parts, and over the large bights, bringing it up again; jam all taut, and it will appear as the sketch.

A RUNNING BOWLINE.

Take the end of the rope round the standing part, through the bight, and make a single bowline upon the running part, and it is done.

A CAT'S-PAW, FOR SETTING UP SHROUDS, ETC.

To form it, lay the end part of the rope or laniard over the standing part, and middle of the bight, then breaking it down, and turning it three times over both parts, and hook the tackle on to both bights.

A COMMON BEND.

Pass the end of a rope through the bight of another rope, or through the becket of a block; then round and underneath the standing part, as shown in the sketch. To prevent it jamming, pass it round twice under the standing part. The sheet of a sail has the end passed up through the clue, then round the clue, and underneath the standing part.

A CARRICK BEND.

This bend is often used in haste, to bend hawsers together, or to form a greater length of warp to tow with. In forming this bend, lay the end of the hawser across its standing part; take the end of the other hawser, and lay it under the first standing part at the cross and over the end; then pass the end down through the bight again on the opposite side from the other end, observing that one end must be on the top, and the other underneath, as is seen in the adjoining sketch.

A FISHERMAN'S BEND.

Take two round turns with the end of a rope or hawser through the ring of an anchor, or round a spar, and one half-hitch around the standing parts, and under all parts of the turns; then one half-hitch around the standing part above all, and stop the end to the standing part; or dispensing with the last half-hitch, tuck the end under one of the round turns, and it becomes a *studding-sail bend.*

A ROLLING BEND.

This is something similar to a fisherman's bend. It is two round turns round a spar, two half-hitches around the standing part, and the end stopped back.

A SELVAGEE STRAP.

A selvagee is used to hook a tackle to any rope, shroud, or stay, to stretch or set up, it being not so likely to slip as a rope strap; two or more turns of the selvagee are taken round

the rope in which the hook of the tackle is fixed.   To make a selvagee strap, get a couple of spike nails and drive them into any convenient place, as far distant as the length intended for the strap; make the end of a ball of rope-yarns fast to one of the spikes, then take it round the other one, and keep passing the rope-yarn round and round in this manner, hauling every turn taut, until it is as stout as it has to be.

When it is to be a very large strap, it is marled down with stout spun-yarn; if of middling size, with two single rope-yarns; and if a small strap, a single rope-yarn.

A PUDDING FOR A MAST OR YARD.

Take a piece of rope of the required length, and splice an eye in each end; put it on a stretch, then worm it, and parcel it with worn canvas according to the shape wanted. By the sketch it will be seen that they are made large in the middle, tapering gradually towards the ends, and made flat on the side which goes next the yard or mast.   When made to the size required, marl it down, beginning in the middle, and marling both ways to the eyes.   If the pudding is for a yard, it is commonly covered with thick leather or green hide; but when for a mast, it is neatly pointed over.

TO FORM AN EYE-SPLICE.

An eye-splice forms an eye or circle at the end of a rope, on itself or round a block.   The strands are first unlayed, and laying the strands at any distance upon the standing part of the rope, according to the size of the eye-splice required, open the lay of the rope with a fid or a marline-spike, and put the middle strand through first, then pass it over the surface of the second strand, and push it through the third; repeat the same with the two other ends, laying them fair apart, observing to taper the strands by gradually reducing the yarns.

AN ARTIFICIAL EYE.

Take the end of a rope and unlay one strand to a certain length, and form the eye by placing the two strands along the

standing part of the rope and stopping them fast
to it; then take the odd strand and cross it over
the standing part, and lay it into the vacant place
which it was taken from at first; work around
the eye, filling up the vacant strand until it comes
out at the crutch again, and lies under the other
two strands; the ends are tapered, scraped down,
marled, and served over with spun-yarn.

THE CUT OR CONT SPLICE.

This is to form an eye in the middle of a rope, as the eye-splice
doth at the end. Cut the rope in two, and unlay the strands
of each; then lay the ends of one
rope on the standing part of the
other, and stick the end through
between the strands similar to an
eye-splice, and do the same with the
other ends, so that the rope becomes
double in the extent of the splice. This splice or collar is
occasionally used for pendants, jib-guys, breast-backstays, odd
shrouds, &c.

A FLEMISH EYE OR MADE EYE.

Unlay the end of a rope, then open the yarns,
divide them into parts, and take a piece of round
wood the size intended to make the eye, and half-
knot about one-half of the inside yarns over the
piece of wood; scrape the remainder down over the
others; then well marl, parcel, and serve them
together. This makes a snug eye for the collars of
stays.

A SHORT SPLICE.

A short splice is made by unlaying the ends of two ropes, or
the two ends of one rope to a sufficient length, then crutch them
together, as per adjoining sketch;
draw them close, and push the
strands of one under the strands of
the other, the same as the eye-splice.
This splice is used for block-straps,
slings, &c. If the ends are to be
served over, they are but once stuck
through; if not, they are stuck twice
and cross-whipped across the strands,
so as to make them more secure. When the ends are to be
served, take a few of the underneath yarns, enough to fill up the

lay of the rope for worming, then scrape or trim the outside ends, and marl them down ready for serving.

A LONG SPLICE.

A long splice is made to rejoin a rope or ropes, intended to reeve through a block, without increasing its size. To make it, unlay the ends of the ropes to a sufficient length, which may be from one half to a whole fathom in length, crutch them together in the same manner as a short splice; one strand is then unlaid, and the opposite strand laid up its intervals; then turn the rope round and lay hold of the two next strands that will come opposite their respective lays; unlay one and fill up with the other as before; the ends are then split equally in two, and the two opposite half strands are knotted together at the ends and middle of the splice, so as to fill up the vacant lay; then stick the ends twice under two strands with all six of the half strands, leaving the other six neutral; the splice is then well stretched before cutting off the ends, and it is finished.

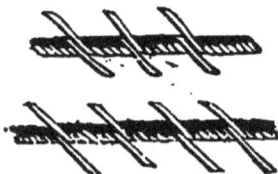

A long splice of four-strand rope is made in a similar way as the preceding.

TO WORM AND SERVE A ROPE.

Worming a rope is to fill up the contlines or vacant space between the strands of the rope with spunyarn or small rope, in order to strengthen it, and to render the surface smooth and fair for parcelling. The first end of worming is securely stopped, and, when arrived at the end of the length intended to be served, it is there stopped, then laid back into the second vacant space; and so on successively, stopping it at the ends.

Parcelling a rope is wrapping old canvas about it, cut in long narrow slips, well tarred and rolled up in rolls before commencing to lay it on the rope. It is customary with some to put on parcelling with the lay of the rope in all cases; but for rigging, which is not intended to be served over, the parcelling ought to be put on the contrary way.

Serving a rope is encircling it with line or spunyarn, &c., to keep it from rubbing and chafing. The end of the spunyarn, for service, is placed under the two or three first turns to keep it fast; then two turns are taken round the mallet and rope, as shown in the sketch. The mallet is then turned round the rope by its handle, while a boy passes

the ball of spunyarn at some distance from the man that is serving the rope, and passes it round as he turns the mallet, until the rope is covered the length required; when the mallet is within a few turns of the end, take the turns off the mallet and pass them by hand, the ball or end is put through under the three or four last turns of the service, and hauled taut, where it is made fast, as at first.

CHAPTER VIII.

To put a Strand in a Rope.—To make a Grommet.—To Sheep-shank a Rope or Back-stay.—To make a Turk's Head.—Wall Knot.—To Wall and Crown.—Shroud Knot.—A French Shroud Knot.—A Matthew Walker. —A Spritsail-sheet Knot.—A Diamond Knot single.—A Diamond Knot double.—Common Sennit.—A Sea Gasket.—A Wrought or Panch Mat. —A Harbour Gasket, or French Sennit.—Pointing a Rope.—A Stopper Knot.—Buoy-rope Knot.—To clap on a Throat and Quarter-seizing.—To pass a Rose-lashing.

TO PUT A STRAND IN A ROPE.

This is done when it happens of one strand of a rope getting chafed or magged, and the other two remaining good. To manage this, cut the strand at the place where it is chafed, and unlay it about two feet each way; then take a strand of a rope about the same size, and lay it in the vacancy of the rope, as shown in the sketch, and stick the ends the same as a long splice, and it is done.

TO MAKE A GROMMET.

A grommet is made by a strand of a rope, and placing one part over the other; with the long end follow the lay, until it forms a ring or small wreath with three parts of the strand all round; finish it by knotting and splicing the ends the same as a long splice.

TO SHEEP-SHANK A ROPE OR BACKSTAY.

This is done to shorten a backstay, when the mast is struck; the rope is doubled in three parts, as shown in

the figure, and taken a hitch over each bight with the standing part of the backstay, and jammed taut.

TO MAKE A TURK'S HEAD. •

To make this, take a round turn round the rope with a piece of log-line, cross the bights on each side of the round turn, and stick one end under one cross and the other under the other cross; it will then be formed like the middle figure of the sketch: after which follow the lead until it shows three parts all round, and it will form the Turk's head.

Turk's heads are generally made on man-ropes, and some times on the foot-ropes of jib-booms, in lieu of an overhanded knot, as they are much neater than the knot, and thought by many seamen an ornament.

WALL-KNOT.—TO WALL AND CROWN.

To form a Single Wall.

Single Wall.

To form a Single Wall and Crown.

Single Wall and Crown.

Walled, Crowned and Walled.

Double Walled and Double Crowned, or Man-rope Knot.

To make the wall, unlay the end of a rope, and with the three strands form a wall knot, by taking the first strand and forming a bight; take the next strand and bring it round the end of the first, the third strand round the second, and up through the bight of the first; this is a wall. To crown this, lay one end over the top of the knot, which call the first, then lay the second over it, the third over the second, and through the bight of the first. It will then appear as the sketch. To *Double Crown:* this is made by unlaying the strands sufficiently, and there making a stop with rope yarn; then single wall and crown, then double wall and double crown, and haul the end tight, and jam the knot: then the strands are led down through the walling, and laid down in the contline; tapered, marled, and served over with spunyarn.

SHROUD KNOT.

Unlay the ends of two ropes about four feet, and interplace one in the other, the same as you commence to make a short splice; then a single wall-knot is made with the ends on each standing part, and the end laid in the contline, tapered down, and served over with spunyarn. This knot is used when a shroud is either shot or carried away.

A FRENCH SHROUD KNOT.

Place the ends of two ropes as the preceding, drawing them tight together; then lay the first three ends back upon their own part, and single wall the other three ends round the bights of the other three and the standing part; it will then appear like the annexed sketch. The ends are tapered as the last. This knot is much neater than the common shroud knot.

A MATTHEW WALKER.

A Matthew Walker is made by separating the strands of a rope, and taking the first strand round the rope and through its own bight; then take the second end round the rope underneath through the bight of the first, and through its own bight; the third end take round the same way, underneath and through the bights of all three. Haul them taut, and they form the knot as the sketch. It is a handsome knot for the end of a laniard, if well made.

A SPRITSAIL-SHEET KNOT.

Unlay two ends of a rope about two feet, and place the two parts which are unlaid together; form a bight with one strand,

and wall the six together against the lay of the rope, the same as was done in a single wall with three ends; after this is walled with the six ends, haul them taut; you must then crown with the six ends, and it will appear as the sketch. To complete it, follow the lead of the parts, and double wall and crown it.

A DIAMOND KNOT SINGLE.

The strands of the rope are unlayed a sufficient length to make the knot; then form bights, by laying the three strands down the sides of the rope, and keep them fast with your left hand; then pass the end of the first strand over the bight of the second strand and through the bight of the third; then take the second over the third and through the bight of the first; then the third over the first and through the second. Haul these taut, and lay the ends of the strands up again to the next knot. These knots are used as ornaments upon bell-ropes, and for jib-boom foot-ropes, man-ropes, &c.

A DIAMOND KNOT DOUBLE.

This is made by the several strands following their respective places through the bights of the single knot, the ends coming out at the top of the knot; lay the ends of the strands up as before.

Common Sennit is braided cordage, made by plaiting from three to any number of rope yarns together, one over the other, according to the size and length, always keeping an odd yarn.

A *Sea Gasket* is made by taking three or four foxes, according to the size required to make the gasket: three or four are plaited together, long enough to make the eye; this being done, clasp both parts together to form the eye over a belaying-pin, and plait it by bringing the outside foxes on each side alternately over to the middle; the outside one is laid with the right hand, and the parts held steadily until the whole is together, adding a fox when necessary. When of a sufficient length to taper, diminish by leaving out a fox at proper intervals. At the finish of it, one end is laid up, allowing enough to form a bight; then plait the others through this bight a few times; the end which was laid up is hauled tight to secure all parts. The ends are cut off, and the end is whipped.

A Wrought or Panch-Mat.—A small rope or line is stretched in a horizontal direction, and made fast at each end, across which foxes (according to the breadth the mat is to be made) are middled and hung over it; then beginning with the first next the left hand and twist a turn in the two parts, and one part give to the man opposite (two men being employed to work the mat); the next fox has a turn twisted in its two parts, and one part given back to the opposite man; the remainder are twisted round the first which are given back, and then again round its own part, and so on in succession. This will make the mat downwards; and, when finished to the length intended, it is begun again at top till its breadth is completed. Each twist is to be pressed tight, and each couple of foxes is to be twisted together at the bottom, to keep in their twists till the next in succession are interwoven with them. At the bottom of the mat selvage it by taking another small rope or line across in a tight manner, similar to the head-line, round which one fox is half-hitched while the next fox is laid up at the back of it, and so on alternately. Trim the ends off, and thrum it with pieces of old strands of rope, cut in pieces about three or four inches long; open the lays of the foxes with a small marline-spike, push the thrums through the lays, and open their ends out.

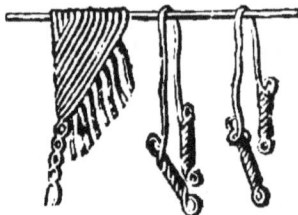

A Harbour Gasket, or French Sennit, is made with foxes, something similar to the common sea gasket; but, instead of taking the outside fox over all the rest, and bringing it into the middle, it is interwoven between them by taking the outside fox of both sides, and taking it over one and under the other, working it towards the middle, the same as common sennit.

Pointing a Rope.—Unlay the end of the rope a sufficient length, and stop it; open the strands out into yarns, and take out as many as it will require to make the knittles,* by splitting the yarns and making one knittle out of every outside yarn; when they are made, stop them back on the standing part of the rope; then form the point with the rest of the yarns, by trimming and scraping them down to a proper size, and marl it down with twine. Divide the knittles, taking every other one up and every other one down; then take a piece of twine, called the warp, and with it pass these turns very tight, taking a hitch with the last turn every time passing the warp or filling. Then take the knittles which are up and bring them down, and the ones which are down, up; hauling them tight, and passing the warp every time over the lower knittles; proceed in this manner until it is got almost to the end, reserving enough of the knittles to finish it with; leave out every other bight of the knittles of the last lay, and pass the warp through the bight, haul them taut and cut them off. A becket is sometimes worked in the end.

A STOPPER KNOT.

This is made by double-walling and crowning, which has been described in page 78. The ends are put up through the heart, and whipped at top.

BUOY-ROPE KNOT.

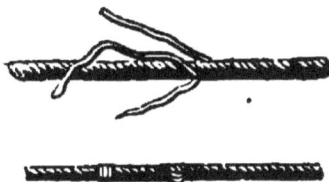

Unlay the strands of a cable-laid rope, take one strand out of the large ones, and then lay the three large ones up again as before; take the three small ones which were left out, single and double them round the standing part of the rope; then take and worm the spare ends along the lay, and stop them.

TO CLAP ON A THROAT AND QUARTER-SEIZING.

To make a round or quarter-seizing,† splice an eye in one end of the seizing, and taking the other end round both parts of the

---

* Knittles are made by laying rope yarns together, with your finger and thumb, against the twist of the yarn.

† Seizing is joining two parts of a rope together with spunyarn, houseline-marline, or small cordage.

rope that the seizing is to be put on; then reeve it through the eye, pass a couple of turns and heave them hand taut; then make a kind of cat's-paw on the seizing by the *marline-spike*, and laying the end over the standing part, push the marline-spike down through, then under the standing part, and up through the bight again. Heave these two turns well taut with the spike, pass the rest and heave them taut in the same manner, making six, eight, or ten turns, according to the size of the rope; then push the end through the last turn, and pass the riding turns five, seven, or nine more (which are termed *riders*), always laying one less of the riding than of the first turns. These are not to be hove too taut, that those underneath may not be separated. The end is now pushed up through the seizing, and two cross-turns taken between the two parts of the rope and round the seizing, leading the end under the last turn, and hove well taut; make an overhanded knot on the end of the seizing, and cut off close to the knot.

When the seizing is put on the end of a rope, and round the standing part, it is called an *End-Seizing*. If on the two parts below the end, a middle or *Quarter-Seizing*.

A *Throat-Seizing* is passed the same way, with riding turns, but not crossed with the end of the seizing. A bight is formed by laying the end over the standing part; the seizing is then clapped on, the end put through the last turn of the riders, and knotted. The end part of the rope is turned up, and fastened to the standing part; this is used for turning-in dead-eyes, hearts, blocks or thimbles.

TO PASS A ROSE-LASHING.

This lashing is passed crossways over and under one eye, then under and over the other; the end part is afterwards taken in a circular form round the crossing, and the end tucked under the last part. This circular part is done to expend the end, instead of cutting it off, so that it will answer again for the same purpose. The use that this is applied to is in lashing a strap or pudding round a mast or yard, or the parral-lashing of a topgallant-yard.

CHAPTER IX.

BLOCKS.—A SHELL, PIN, AND SHEAVE.

BLOCKS are used for various purposes in a ship, either to
increase the mechanical power of the rope, or to arrange the
ends of them in certain places on the deck; and they may be
readily found when wanted: they are consequently of various
sizes and power, and obtain various names according to their
form or situation.

Every block consists of three, and generally four, parts :—
1. The shell, or outside wooden or iron part. 2. The sheave, or
wheel, on which the rope runs. 3. The pin, or axle, on which
the sheave turns. 4. The strop, or part by which the block is
made fast to any particular station, and is usually made either
of rope or of iron. *Iron-stropped Blocks* frequently have the
hook working in a swivel in order to turn it, that the several
parts of the rope of which the tackle is composed may not be
twisted round each other, which would greatly diminish the
mechanical power.

The shell of a block is made of ash, elm, or iron, and
has one or two scores cut at each end, according to its
size; these scores are for the purpose of admitting a
strap, which goes round the block, in the centre of which
is a hole for the pin; the shell is hollow inside to admit
the sheave.

The sheave is a solid wheel, made of lignum vitæ, iron, or
brass; in the centre is a hole for the pin, on which it turns.

The lignum vitæ sheave is bushed with brass or iron;
round the circumference a groove is cut, that the rope
which goes over it may play with ease. The sheave is
placed in the shell, and the pin is put through both shell
and sheave, which constitute a block.

NAMES OF THE DIFFERENT BLOCKS, AND THEIR USES.

What is termed a single block has but one sheave, and if intended for a double strap there are two scores on the outside of the shell. Single blocks are more used than any other kind on board of a ship.

A double block has an additional sheave; it is otherwise the same as a single block.

A treble block is made in the same manner as a double, with one more sheave. Treble blocks are generally used as purchase blocks.

A Shoulder block is the same as a single block, with the exception that it has a projection at the bottom of the shell, called a shoulder, to prevent the rope that reeves through it from jamming between the block and the yard. These blocks are mostly used for bumkin or lift-blocks on lower yards.

A Fiddle block is made like two single blocks, one above the other; the upper one being the largest, so as to allow the rope which is rove in the upper sheave to play clear of the rope in the under one. These blocks are used in places where there is not space enough for a double one, or where it (the double block) would be liable to split by not "canting" fair, or having room to play. These blocks are used for top burtons, &c.

A Shoe block is also made like two single blocks, but the sheave of the upper one lies in a contrary direction to that of the lower one. They are generally used as buntline blocks to courses; the buntline reeving in the upper sheave, and the whip in the lower one.

A Sister block has two sheave holes one above the other; three scores for seizings, one at each end, and one between both sheaves; they are hollowed out on each side of the shell to take the shroud. These blocks are used as topsail lift and reef-tackle blocks, and are seized-in between the two forward shrouds of the topmast rigging, above the futtock stave. The lift reeves through the lower sheave, and the topsail reef-tackle through the upper one.

A Dead-eye is a large round piece of wood with three holes in it, and a groove cut round it for the shroud to lie in. It is used to turn in the ends of shrouds and backstays; the three holes are used to reeve the rope or laniard through, which is well greased to reduce the friction when setting up the shroud or backstay. The round shape of the block, and the position of the three holes,

give it somewhat the shape of a death's head, and hence its name, "the Dead-eye."

A *Bull's-eye* is a kind of thick wooden thimble, with a hole in the centre, and a groove cut round the outside for the rope or seizing to lay in.

A *Heart* is a peculiar sort of dead-eye, resembling a heart; it has one large hole in the centre, at the bottom of which are four or five scores, and round the outside is a groove cut to admit a rope called a stay. There are other hearts, called "collar-hearts," which are open at the lower ends, opposite to which the laniard is passed. This heart has a double score cut round the outside, and two grooves cut on each side for the seizing to lay in, which keeps the collar in the scores of the heart. Hearts intended for bob-stays should be made of lignum vitæ; those made of ash being liable to split.

A *Belaying-Pin Rack* is a piece of wood with a number of holes through it, in which belaying-pins are stuck; on the back part are several scores for the shrouds to lie in, to which it is seized.

A *Euphroe* is a long piece of wood, having a number of holes, through which the legs of the crowfoot is rove; a score is cut round it to admit a strap. This is used for the ridge of an awning.

Ninepin Block.

Monkey Block.

STRAPPING OF BLOCKS.

The whole length of all the different sizes of block-strapping is got upon the stretch, and hove out taut for worming and serving; it is then wormed and served, and the required number cut into lengths to suit the different blocks. A common strap is fitted in the following manner:—First cut the rope once-and-a-half the round of the block, then get it on a stretch; worm, parcel, and serve as near the end as possible, not to interfere with splicing; then splice the ends together with a short-splice, and finish serving snug up to the splice. Stretch it and cut the ends off, or serve over the ends.

TABLE OF THE DIMENSIONS OF STRAPS AND SEIZINGS FOR SINGLE AND DOUBLE BLOCKS.

Size of Blocks.	Size of Strap.	Length when spliced for Single Blocks.	Seizing for Single Blocks.	Length when spliced for Double Blocks.	Seizing for Double Blocks.
Inches.	Inches.	Ft. In.	Marline.	Ft. In.	Marline.
5	1½	1 5	6 foot	1 7	6 foot
6	2	1 6	6 ,,	1 9	6 ,,
7	2	1 0	7 ,,	2 0	7 ,,
8	2½	2 0	9 ,,	2 3	10 ,,
9	3	2 3	11 ,,	3 0	13 ,,
10	3	3 0	13 ,,	3 3	15 ,,
			Rope. Ins. Yms.		Rope. Ins. Yms.
11	3½	3 3	½ 2½	3 6	½ 3
12	4	3 6	½ 3	3 9	¾ 3½
13	4	3 0	¾ 3½	4 3	¾ 3½
14	4½	4 2	1 3½	4 6	1 3½
15	5	4 5	1¼ 3½	4 9	1¼ 4
16	5½	4 8	1¼ 4	5 1	1¼ 4
17	6	5 1	1½ 4	5 7	1½ 4
18	6½	5 7	1½ 4	6 2	1½ 4
19	7	6 1	1½ 4	6 9	1½ 4
20	7½	6 9	2 3½	7 4	2 3½

NOTE.—In cutting straps from the 3-inch rope upwards, 18 inches more length will be required for splicing, &c.; under 3-inch, 12 to 15 inches.

Blocks strapped with eyes or thimbles in the ends, are seized tight into the bight, and the legs left long enough to lash through the eyes, round the mast, yard, &c., as the topsail clue-lines, clue-garnets, &c. Girt-line blocks are strapped with a lashing eye or tail, and the girtline rove. Blocks, strapped with double tails, are fixed in the strap, similar to blocks with eye-straps; and those with a single tail, called—

A TAIL-BLOCK.

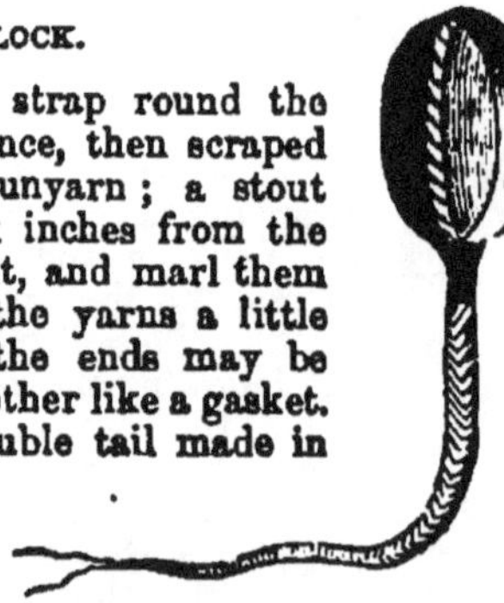

An eye-splice is made in the strap round the block; the ends are stuck but once, then scraped down, and served over with spunyarn; a stout whipping is clapped on about six inches from the splice. Then open the strands out, and marl them down selvagee fashion, tapering the yarns a little towards the end of the tail; or, the ends may be twisted into foxes, and platted together like a gasket. Blocks used for jiggers have a double tail made in the same manner.

A PURCHASE-BLOCK.

This block is double strapped, having two scores in the shell for that purpose; the strap is wormed, parcelled, and served, or only wormed and parcelled, and spliced together. It is then doubled so as to bring the splice at the bottom of the block. The seizing is put on the same way as any other; the only difference is that it is crossed both ways through the double parts of the strap. These block-straps are so large and stiff, that it requires a purchase to set them securely in the scores of the block, and bring them into their proper place.

A TOP-BLOCK.

This is a single iron-bound hook-block; it hooks to an eye-bolt in the cap. The top pendants are rove through the top-blocks when swaying up or lowering down the topmasts.

A CAT-BLOCK.

The cat-block is a two or three-fold block, iron-bound, with a large iron hook attached to it, and is employed to draw the anchor up to the cat-head. On the forward side of the shell of this block are two small eye-bolts, for the purpose of fitting a small rope, called the back-rope bridle, used in hooking the cat.

A SNATCH-BLOCK.

A snatch block is a single block, iron-bound with a swivel hook. An iron clasp is fitted on the iron band or strap, with a hinge to go over the opening or snatch, and toggles on the opposite side. The bight of a rope or a hawser is placed in this block when warping the ship, &c., instead of reeving the end through, which, in some circumstances, would be very inconvenient. Blocks of this description, and of a large size, are generally termed "viol, or rouse-about blocks."

THE SPRING-BLOCK.

The spring-block is an invention of Hopkinson, of Philadelphia, calculated to assist a vessel in sailing, and particularly intended by him to be applied to the sheets and the dead-eyes; it is composed of a common block or dead-eye, attached to a spiral spring of well-tempered steel, within the cavity of which is a chain of suitable strength, called a check-chain; when the spring is not in action, this chain is slack; but when the spring is extended by the force of the wind as far as it may be without injury, the check-chain begins to bear, and prevents its farther extension.

A SINGLE WHIP.

A single whip is the smallest and most simple purchase in use. It is made by reeving a rope through a single block, as the annexed sketch. It is used to hoist up light bodies out of the hold, such as empty casks, &c.

WHIP AND RUNNER.

A rope rove through a single block is called a whip as above; and if the fall of this whip be spliced round the block of another whip, it becomes whip on whip, or whip and runner. Thus two single blocks will afford the same purchase as a tackle having a double and a single block, and with much less friction. To topsail and topgallant yards that hoist with a single tie, there is sufficient length of the hoist to apply the purchase as halliards, which will overhaul with great facility.

A GUN-TACKLE PURCHASE.

This purchase is made by reeving a rope through a single block, then through another single block, and make the end fast to the one it was first rove through, or splice it into the bottom of the block for neatness.

A LUFF-TACKLE PURCHASE.

Luff-tackles are composed of double and single blocks, strapped with a hook and thimble; the rope is rove through one of the sheave holes of the double block, then through the single one, through the double one again, and the end made fast to the single one, with a becket bend, to a becket in the bottom of the block.

A TOP BURTON TACKLE.

This is rove in the same manner as a luff-tackle purchase; the only difference is that the upper block of the burton is a fiddle block, while that of the luff is a double one.

A RUNNER AND TACKLE.

A runner tackle is the same purchase as a luff-tackle applied to a runner. A runner is a thick rope rove through a single block, and has usually a hook attached to one of its ends, and one of the tackle blocks to the other: in applying it, the hook of the runner, as well as the lower block of the tackle, is fixed to the object intended to be removed.

A LONG TACKLE.

A long tackle is composed of two blocks; a long tackle block is double, but it resembles two single blocks joined together at their ends.

A Two-fold Purchase consists of two double blocks; the fall is first rove through one sheave of the upper block; then through one of the lower ones; through the upper one again, then through the lower one, and make the end fast to the upper block.

A Three-fold Purchase is rove in this way: the blocks having one more sheave, commence to reeve the fall in the middle sheave first, instead of one of the side ones, which brings a cross in the fall. The reason of its being rove in this manner is that the heaviest strain comes first on the fall part, and if it was rove in the side sheaves it would have a tendency to *cant* the block in the strap, split the shell of the block, and cut the fall; but when it is in the middle sheave it draws all down square alike.

CHAPTER X.

—◆—

DRAWING PLANS FOR CUTTING RIGGING.

THE most proper way to ascertain the lengths of all standing and running rigging, is to make a *draft*, or rigging plan of the vessel you are employed upon, drawing it to a scale of reduced proportion to the real dimensions, as the 8th or 4th of an inch to the foot, as may be convenient for the drawing.

To draw the plan of rigging for a new ship, it is necessary to have the dimensions of the hull, as :—

The distance between the foreside of the stem to the centre of the foremast.

The distance between the centre of the foremast to the centre of the mainmast.

The distance between the centre of the mainmast to the centre of the mizenmast.

The distance between the centre of the mizenmast to the outside of the taffrail.

The housing of the foremast.

  ,, ,, mainmast.

  ,, ,, mizenmast.

The step of the foremast above a straight line from the step of the mainmast.

The step of the mizenmast ditto ditto.

The number of inches the foremast rakes to the foot.

  ,, ,, mainmast ,,

  ,, ,, mizenmast ,,

  ,, ,, bowsprit rises to the foot.

The height of the rail or gunwale.

  ,, ,, topgallant forecastle.

  ,, ,, poop.

  ,, ,, cathead or bumkins.

Also, the dimensions of masts, yards, gaffs, &c.

[See the adjoining plate.]

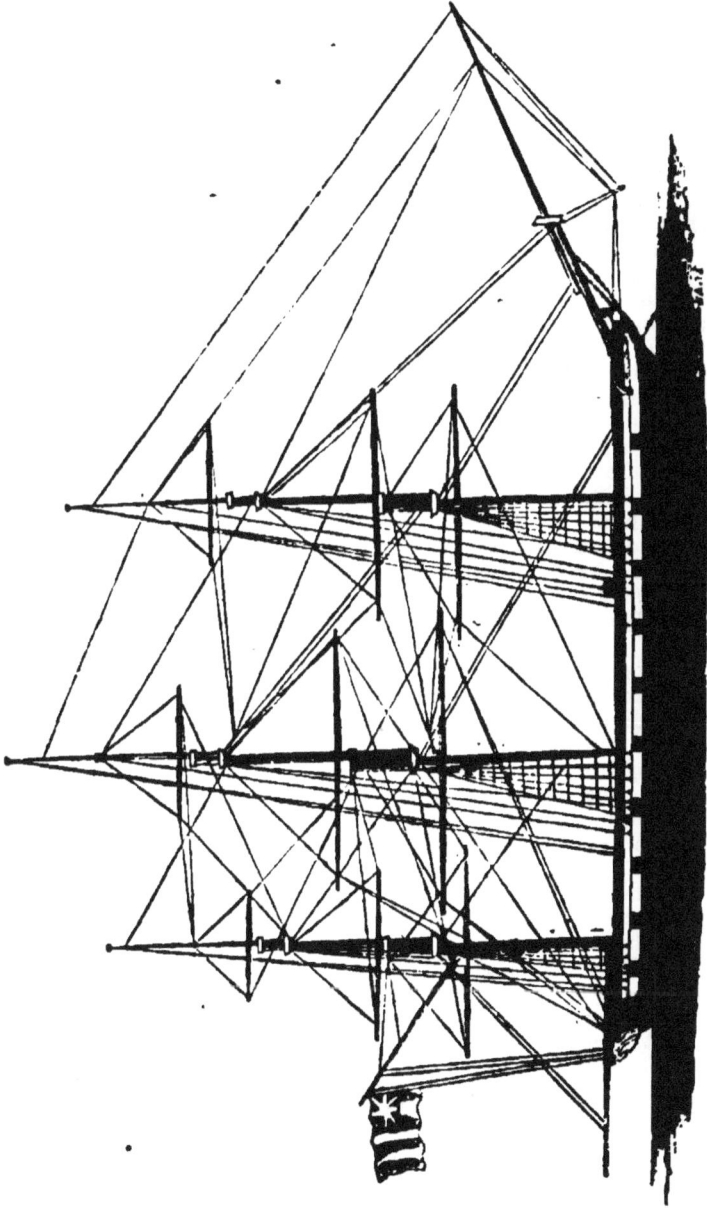

DRAWING A RIGGING PLAN FOR SHROUDS.

Lower Shrouds.— For the length of the shrouds of lower rigging, draw the breadth of the ship from outside of the channels to outside of the channels, from the same scale as the sheer or broadside plan is drawn; set up the height of the masts above the deck to the hounds, and diameters of them. Draw the rigging as the adjoining sketch; then will the distance from the larboard side of the mast-head to the foremost dead-eye in the starboard channels, be the length of the *first pair* of shrouds, making due allowance for the size of the dead-eyes and for stretching in setting up.

As the shrouds spread to channels, which are placed aft of each mast, as shown in the *plate*, their respective lengths are ascertained by applying the length of the shrouds of the foremost ones on this *draft*, which is a guide for each shroud of the carry-aft. Measure them in the same manner as those in the annexed sketch; but allowing for each pair of shrouds to lap over the diameter of the

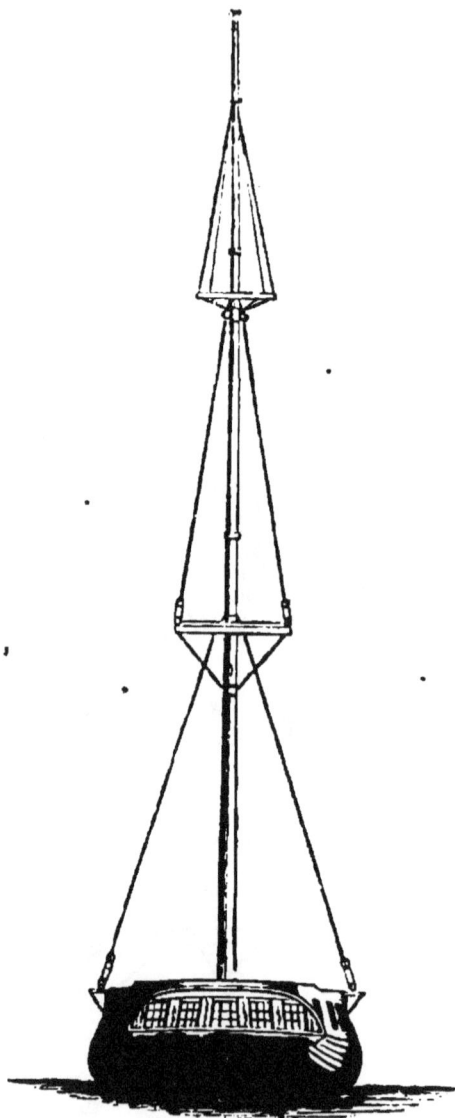

rope at the eye on the mast-head. The length of the shrouds must vary inversely as to the rake of the mast. The greater the rake, the shorter the aftermost shrouds.

Topmast and Topgallant Rigging.—The lengths are found in the same way.

In measuring the length of the shrouds, some prefer the distance from the opposite side of the mast-head to the partners, added to half the breadth of the deck, from the mast to the side.

CUTTING-OUT STANDING RIGGING.

Lower Rigging.—When the rigging plan is completed and the length measured, get the shroud warp on a stretch, or rather one end of it, long enough for one pair of shrouds; mark off the distance for the required service; and, when completed, being wormed, parcelled, and served, on a taut stretch for a few days (the longer time the better), measure the length with a tape-line, of first pair of shrouds, No. 1, starboard; when measured and chalked the required length, slacken down the stretch, and cut at the chalk-mark; middle the shroud at the centre of the service, and lay it on the rigging-loft floor for turning in the dead-eyes, &c. Continue fitting, and cutting, from the *draft*, in this way until is cut the number of shrouds required for the gang, allowing each pair of shrouds to lap over the diameter of the rope at the eye, as they are laid on the loft-floor; alternately making due allowance at the ends, before cutting, for the carry-aft, or the jump of a port, if required. But the exact length of each is easily got from the rigging plan.

In worming, start before the shrouds are hove out to lengthen, because the worming increases in tension with the rope, and thereby draws smooth and even into the outline. In parcelling, begin at each end where the service is to leave off, and parcel upwards to the middle of the eye, where commence serving downwards on each leg. The eye-seizings are round ones, and when put on, the whole eye is neatly covered with parcelling. A half-sister block is sometimes put on between the two forward shrouds, for the lower boom topping lift.

Top-mast and Topgallant Rigging is cut in the same manner. In fitting the top-mast rigging, always seize in a sister-block between the two forward shrouds, for the top-sail lift and reef-tackle. The swifters are generally served the whole length.

The eyes of the topgallant rigging are made to fit exactly around the cylinder; if there is an odd top-mast shroud, or topgallant-shroud, on each side, they are either fitted with a horse-shoe eye, or go together with a cut splice.

BACKSTAYS.

Breast and Standing, are stays which support the top-mast, topgallant, and royal-masts from aft; they reach from the heads of their respective masts to the channels on each side of the ship, and assist the shrouds when strained by a press of sail, as shown in the plate, p. 92.

These may be cut by the same rule; the eyes of the breast backstays are fitted in different ways. They are sometimes spanned together, making a square, the size of the mast-head; sometimes they have an eye like the shrouds, made to fit close; and others have a small eye seized in the bight, and lashed round the mast-head. The eyes of the standing backstays are fitted like those of the shrouds.

FORE AND AFT-STAYS.

These being marked on the rigging plan (see plate, p. 92), measure from the after parts of the mast-head to where the stays set up, and to this distance add the length of the mast-head, for collars.

Collars for stays are the length of their respective mast-heads. The mousings are raised once and a half the size of the stays, and at a distance equal to twice the length of the mast-head from the mousing. A Flemish eye is worked on the end, and the stay rove through it; or they may be fitted with lashing eyes, in which case each leg is the length of the mast-head; the service is continued the length of the eye below the mousing, the collars parcelled or leathered, and the hearts turned in with the lay of the rope. Stays are wormed, parcelled, served, and leathered in the wake of all nipps, such as the bees, bullock-blocks, and sheave-holes.

CUTTING LOWER MAST-HEAD PENDANTS.

The forward pair should be twice the length of the mast-head,—the after pair twice and a half; thimbles are spliced in the ends, and they are wormed or spanned together, so as to form a span to fit the mast-head.

PUTTOCK SHROUDS.

The distance from the extremity of the top to the puttock-hoop, or chain-necklace, will give the length of the puttock-shrouds, which must have a hook and thimble in their upper ends, and a thimble in their lower ends. The puttock shrouds are hooked to their respective plates in the tops, *with the points of the hooks in.* See sketch, p. 13.

BOBSTAYS.

The bobstays of all merchant vessels are chain, which are fitted with shackles to the cutwater, with iron plates let in flush with the wood, a bolt going through both plates; the other ends are set tight by screws to the hoops on the bowsprit. See p. 20.

BOWSPRIT SHROUDS.

These are of chain, and the length may be found by making an athwartship plan of the rigging of the bowsprit. A dead-eye or heart is attached to the end which sets up to the collar or hoop on the bowsprit, and a hook at the other, to hook to the eye-bolts in the bows.

JIB AND FLYING-JIB GUYS.

Take the distance from the boom-end to the spritsail-yard-arm, and from thence to the bows, if an athwartship *draft* be made ; or make an allowance for reeving through the spritsail-yard. They are generally fitted with a cuckold's-neck over the boom end, and set up with dead-eyes to the bows.

When no spritsail-yard is carried, the jibboom is secured by guys to the outriggers commonly called *whiskers*, which are placed just inside the bowsprit caps ; but sometimes extend out from the fore part of the cat-heads, and in this case made of iron, with sheaves at the extremity, through which the jib-guys lead, and are set up inboard.

The martingales must be cut, and fitted to the manner in which they are rove.

TURNING-IN DEAD-EYES.

Take the length of the shroud from the draft, if the masts are not stepped, and place the dead-eye to that length, as directed in p. 93. Turn the dead-eye in as near the end as possible, so that all parts of the shroud may be equally stretched, observing to keep the lay in the rope, as it prevents the wet getting in. The score of the dead-eye being well tarred, is thus turned in,—the end of the shroud is taken underneath round the dead-eye, inside standing, or mast-head part ; a bolt is put in a hole of the dead-eye. The dead-eye machine (a pair of screws) is fixed on, and the shroud is hove in quite snug round the dead-eye.

When the shroud is hove well round, pass a good throat-seizing ; when secured, bring the end taut up ; then pass a round, or quarter seizing, and a smaller one on the end.

There is one thing of importance, and should be observed in turning rigging-in on shore—to keep the lay in the rope, and when sent out of the loft, to be placed on the mast-head, keep the ends inside, the shrouds being marked with a knot or a piece of spunyarn, according to the number. The ends will lay aft on one side, and forward on the other.

Turning in dead-eyes, termed *Cutter-stay fashion.*—The dead-eye being placed to the mark, the end is passed round it as before, but instead of being secured with a throat-seizing, the end is passed round the standing-part, and seized to the part round the dead-eye with a round-seizing, and another on the end round the dead-eye.

CHAPTER XI.

Rigging Sheers.—To take in the Mizen-mast.—To take in the Main and Fore-mast.—To take in the Bowsprit.—Gammoning the Bowsprit.—Rigging the Fore, Main, and Mizen-masts.—Lower Tackle—Pendants, Shrouds, Swifters, Stays, the setting up of the Lower Rigging.—Rigging the Bowsprit—Bobstays, Bowsprit Shrouds, Blocks for the Fore-bowline, Blocks for the Foretop-bowline, Horses or Ridge-ropes, the Goblines or Back-ropes.—Getting the Tops over.—Rigging the Top-mast—Getting the Cap into the Top, the Top Tackle Falls and Blocks, getting Top-mast Cross-trees over, Ginn-blocks, placing Top-mast Rigging, to seize in the Sister Blocks, the Top-mast Cap, the Standing After Back-stays, the Fore Top-mast-stays, Main Topmast-stay, Main Topmast Spring-stay, Mizen Top-mast-stay.

RIGGING SHEERS.

EVERY facility is afforded at Her Majesty's dock-yards for lifting the lower masts and the heavier parts of rigging on board, by large "Sheers;" a detailed description of which is given in p. 84, "Rudimentary Construction of Cranes," by Joseph Glynn, Esq, F.R.S. As merchant ships have not recourse to these, and it is only a few places else have got them, as in the East India Docks, London; the new docks at Sunderland, &c.; it becomes necessary to get such spars as can be procured, and erect a pair of sheers on board for that purpose.

In doing this proceed as follows:—Take in a sufficiency of ballast, or coals, to steady the ship, if tender, and shore the decks from the skin up, particularly abreast of the partners. Sling "skids" up and down the sides, for the purpose of keeping the sheer legs clear of the channels; reeve the "par-buckles," (see p. 20, "Construction of Cranes,") and bring the sheer legs alongside with their small ends aft; parbuckle them on board, and their heads or after ends resting either on the taffrail, the break of the poop, or a spar placed in the most

r

convenient spot, the more elevated the better. Square the heels
exactly one with the other, so that when they come to be
raised the legs may be found of equal height.

As near the after ends of the spars as may be considered
necessary, when crossed, put on the head-lashing of new, well-
stretched rope (figure of 8 fashion), similar to a racking-seizing,
and cross with the ends. Open out the heels, carrying one over
to each gangway, and placing it on a solid piece of oak or shoe,
previously prepared for the purpose. Clap stout tackles on the
heels, two on each, one leading forward, the other aft ; set taut
the after ones, and belay them. Lash a three or four-fold
block, as the upper one of the main purchase, over the main-
lashing (so that it will hang plumb under the cross), with canvas
underneath to prevent chafing ; and in such a manner that
one-half the turns of the lashing may go over each horn of
the sheers, and divide the strain equally ; also sufficiently long
to secure the free action of the block. Lash the small purchase-
block on the after horn of the sheers, sufficiently high for the
falls to play clear of each other, and a girtline-block above all.

Middle a couple of hawsers, and clove-hitch them over the
sheer-heads—having two ends leading forward, and two abaft,
led through vial blocks, and stout-luffs clapped on them. These
should be sufficiently strong to secure the sheers while lifting
the masts. The lower purchase block is lashed forward, round
the knight-heads (perhaps round the cut-water), and the fall
being rove, the sheers are raised by heaving upon it, and
preventing the heels from slipping forward, by means of the
heel-tackles previously mentioned.

Sometimes a small pair of sheers are erected for the purpose
of raising the heads of the large ones.

When the sheers are up, the heels confined to their "shoes,"
they can then be transported along the deck by means of the
heel-tackles and guys to the situation required, taking care to
make them rest upon a beam, and to have the deck properly
shored up below.

Finally, give the sheers the necessary rake by means of the
guys, and set taut all the guys and heel-tackles, &c. ; this being
done, the sheers may be considered ready.

TO TAKE IN THE MIZEN-MAST.

The mizen-mast being alongside, with the head aft, and the
garland* lashed on to the forward part of the mast at the
distance from the tenon to just above the spar-deck partners ;

* Garlands are made of new rope, well stretched (selvagee fashion). they
are laid on the forward part of the mast, a stout lashing put on over all, and
crossed between the garland and the mast ; a good dogging also, if necessary,
passed downward.

lash a pair of girtline-blocks on the mast-head, and reeve the girtlines; bend the sheer-head girtline to the mast below the bibbs, to "cant" it. Overhaul the main purchase down abaft, thrust the strap through the eyes of the garland, toggle it, and secure the toggle by a back lashing. Take the fall to the capstan and "heave round;" when the heel rises near the rail, hook on a heel-tackle to ease it inboard. Get the mast fair for lowering by means of the girtlines, wipe the tenon dry, and white lead, or tar both it and the step, "lower away," and step the mast.

Some distance may be saved, by using no garlands and having the purchase-blocks lashed to the mast.

The mast being stepped, and wedged temporarily, "come up" the purchases, man the guy and heel-tackles, and transport the sheers forward for taking in the main-mast.

TO TAKE IN THE MAIN AND FORE-MAST.

Proceed in the same manner as in getting in the mizen-mast. It is better not to use garlands, when the sheer legs are rather short, as lashing the purchase-blocks to the mast shortens the distance. If the ship has a topgallant-forecastle, it would be well to step the mast forward of the sheer legs, for the brake of the forecastle comes abreast of the partners ; and, in a case of this kind, it would be well to take in the foremast first.

TO TAKE IN THE BOWSPRIT.

Transport the sheers as far forward as possible, or as the bows will permit ; send a man to the sheer-head, bend on the girtlines to the small purchase-block to light it up, unlash it, and lash it again to the forward fork or horns of the sheers, pass a strap round the fore-mast head, to which hook a large tackle, carry it well aft ; and haul it taut, for the purpose of staying the mast. Lash a couple of large single blocks to the foremost head, middle, and hawser, and clove-hitch it over the sheer-head ; reeve the ends through the blocks at the mast-head, down on deck, carry them well aft, and take a turn. Hook the after-heel tackles forward, and take the after-guys aft ; pass a bulwark lashing round each heel, rake the sheers over the bows sufficiently for the main purchase to hang directly over the gammoning-scuttle, and make all fast.

The bowsprit being brought under the bows, with the head forward, and the garlands lashed on, the main one a little more than one-third from the heel, the smaller one between the cap and bees, having guys leading from the bowsprit to the cat-heads, and a couple of straps round the heel for hooking the bedding tackles ; overhaul down the purchases and toggle them ; "sway away," attending it by the guys, until nearly perpendicular ; hook on the bedding tackles, which are taken from the bitts on the main-deck, and led up through the partners ; wipe the tenon dry, and white-lead, or tar both it and the mortice ; "lower away," bouse upon the bedding-tackles, and bring it into its place ; come up purchases, guys, unlash garlands, and proceed to dismantle the sheers.

If the ship has a topgallant-forecastle, the bowsprit cannot be taken in with the sheers without the assistance of a *derrick*, on account of the brake of the forecastle, it not being prudent to step sheers on the top of it.

When the ship is masted, and alongside the yard, commence getting on board tops, caps, cross-trees, top-masts, and topgallant-masts ; also have ready tackles and luffs for setting up the rigging and staying the masts, top-blocks, with lashings for top-ropes, and all the rigging at hand and in order.[*]

* See "The Kedge Anchor ; or, Young Sailor's Assistant," by William Brady, of the U. S. *Navy;* in reference to which, I have much pleasure in acknowledging the use I have here made of several articles in his most unique and useful book.

When extreme expedition is not wanted, the following is the usual progressive method of rigging ships:

GAMMONING THE BOWSPRIT.

It is necessary that a stage should be rigged under the bow sprit for this purpose, and slung from the bowsprit end. The gammoning is of new, well-stretched rope; chains generally in the merchant service. One end is passed over the bowsprit, and through a hole cut in the knee of the head, alternately. The first end, if rope, is whipped and passed through the hole, and over the bowsprit, with a round turn, then clinched round the bowsprit close against the cleats or stop; the other end passes through the forepart of the hole, again round the bowsprit, but before the clinch on the bowsprit and aft in the hole. All the succeeding turns go in the same way. A selvagee, or lashing, is put round the cutwater, to which a block is hooked abreast of the hawse-hole, through which a pendant is led through the block, with an eye in its outer end, to which a bight of the gammoning is toggled every turn, while to the other end is hooked a long tackle, and the fall led to the capstan. When all the turns are passed and hove taut, they are frapped together by as many cross-turns as are passed on the bowsprit, each turn hove tight. The end is then whipped and seized in one of the turns. Iron gammoning is put on in a similar way.

RIGGING THE FORE, MAIN, AND MIZEN-MASTS.

Before the trestle-trees are sent up, white-lead the mast-head in the wake of them; overhaul down the girtlines and bend on the trestle-trees, with the after chock out; "sway away;" when above the bibbs slip the stops so as to let them come down gradually into their places; then the after chock is sent up, let in and bolted. Tar the mast-head in the way of the rigging; overhaul again the girtlines for the bolsters, which' are covered with well-tarred canvas; sway them aloft and stop them. The girtline-blocks are now lashed to the after part of the trestle-trees.

The girtlines that reeve through them lead down upon the deck, for hoisting the rigging in the following manner:—

Lower Tackle Pendants.—These are sent aloft and placed, they have got a long and short leg, fitted together with a span, or square, the size of the mast-head. As soon as the mast-head pendants are placed they ought to be lashed abaft, the tackles hooked, and the mast stayed by them.

Shrouds are hoisted over the mast-head. Thus,—Overhaul down the girtlines, bend the mast-head one on the shroud, with a timber-hitch, or toggle, four or five feet below the seizing and stop it to the centre of the eye ; take the girtline from the after trestle-tree, and bend it half way down the shroud ; then "sway away" on the lower girtlines, and lift the weight of the shroud. When high enough, the stop in the eye is cut, and it will fall over the mast-head ; the persons employed to place the rigging laying it fair on the bolsters, beating it well down, and observing to have the eye-seizing come as near the centre of the mast-head as possible. In this manner, hoist foremost pair of shrouds, starboard-side, the next pair forward on the port side ; and so on, alternately, until all the shrouds are over.

Swifters are swayed over the mast-head, next above the shrouds (the after swifter goes over first in small vessels), and are fixed on the starboard and port sides of the mast. In staying the mast these swifters should be set taut, the mast being previously wedged, and the stays set steadily up.

The *Stay* is next sent up, and last of all the *Preventer or Spring-stay.*

The Setting-up of the Lower Rigging.—Reeve the end of the lanyard, if prepared, through the hole of the upper dead-eye nearest to the end, and stopped with a wall-knot, to prevent its slipping ; the other end is passed through the hole of the lower dead-eye, and returning upwards, is rove through the middle hole in the upper dead-eye, and next through the middle hole of the lower dead-eye, and lastly, through the foremost hole in both dead-eyes. Clap a selvagee-strap on the shroud well up, to this hook the single block of a luff-tackle ; the double-block, to a blackwall-hitch in the lanyard ; the fall is then made fast to the hook of the main tackle with another cat's-paw or blackwall ; reeve the tackle fall through the leading block, and pull up, the lanyard being well greased, to make the whole slide with ease through the holes in the dead-eyes. When the rigging is set up for a full due, (which is when the masts are stayed forward and the stays all set up,) the lanyard is first nipped, or stopped, and the end passed between the throat-seizing and the dead-eye with a hitch, then brought round all the parts in turns to expend the lanyard, and the end is well stopped to its own part with spunyarn. The ends of the shrouds are then cut square and capped, and the mats laced on.

RIGGING THE BOWSPRIT.

Bobstays.—Chain is generally used in the merchant service, and fitted to shackle to the cutwater, with iron plates let in flush with the wood, a bolt going through both plates, which is very snug and strong. A heart, or iron-bound dead-eye, is attached to the outer-end, and a lanyard then passes through and connects with the heart, or iron-bound dead-eye, in the iron collar under the bowsprit, and sets up taut, with luff-tackle upon luff, and leads in upon the forecastle. Ships have two or three bobstays, according to their size. Their uses are to bind down and keep steady the bowsprit, and counteract the force of the stays of the foremast, which draw it upwards.

Bowsprit shrouds are single pieces of chain hooked to an eye-bolt on each side of the bow; the foremost end has a heart or iron bound dead-eye linked on; the shrouds are then set taut as the bobstays.

Blocks for the fore-bowlines are spliced, one on each side of the forestay.

Blocks for the fore-top-bowlines are seized, one on each side, to an eye-bolt in the bowsprit-cap.

Horses, or Ridge-ropes.—The outer ends are spliced round a thimble in an eye-bolt on each side of the upper part of the bowsprit cap. The inner ends have a thimble seized in that sets up with a lanyard to an eye-bolt in the knight-heads or stanchions for the purpose.

The goblines or *back-ropes,* whether rope or chain, are fitted to the end of the dolphin-striker, and set up to the bows, one on each side.

GETTING THE TOPS OVER.

The girtlines are overhauled for the cross-trees, are then hoisted into their places, and bolted in the trestle-trees, which are screw-nutted or fore-locked underneath. The top is hoisted on board by the girtlines, and placed up against the aft-side of the mast, except the mizen, which is placed on the fore-side. The girtlines being on each side of the mast-head are then over-hauled; one end is passed from underneath, and up through the hole for puttock-plates; hitch it to the standing part, and stop it with spunyarn through the hole made for the slings in the fore-part, except the mizen-top, which is stopped at the aft-part. A girtline is taken from the mizen-mast head, and bent to the foremost part of the maintop; bend on a tripping-line to the pigeon-hole leading from the foremost-head. The top is then hoisted by its girtlines over the mast-head; when suffi-ciently high to allow the foremost edge of the lubber's-hole to clear the mast-head, cut the stops and cant it over by the tripping-line, and the top will hang in the girtlines, when it can be lowered, placed in its berth, and bolted.

The dead-eyes for the topmast-rigging can now be hauled up, and put in their places in the top-rims, and also ship the top-rail, and puttock shrouds in their respective place.

The top-blocks are large single blocks having iron straps, which are formed, after being put round the block, into a large hook (see sketch, p. 88). Overhaul down the girtlines through the lubber's-hole; then bend one part through the sheave-hole of the block, and stop it to the back part of the hook. The block

is then hoisted up, and lash it to the mast-head around the
hook, with a lashing long enough to allow the block to hang
half-mast-head high. Through this block reeve a hawser, and
send the foremost end down through the square hole in the foremost part of the trestle-trees, the after end through the lubber's-
hole, through a leading block on deck, and round the capstan.

RIGGING THE TOPMAST.

The hawser being already rove, reeve the foremost end through
the sheave-hole in the heel of the topmast, when it is racked to
the topmast in two or three places, between the heel and the
hounds ; it is then well stopped with a good lashing, and enough
of the end to spare to make fast round the mast-head. The
other end is taken to the capstan, through a leading block on
the deck, and the mast hove up. When the topmast is hove
high enough to enter the trestle-trees, the end of the hawser
is clinched round the mast-head and the rackings are cut, the
men in the top being ready to overhaul the girtlines down before
all, and get the cap into the top.

GETTING THE CAP INTO THE TOP.

Reeve the foremost end of the girtline through the round-
hole in the cap, and take two half-hitches ; stop the girtlines
along to the after part of the square-hole. Sway up the cap
until it is high enough to clear the forepart of the top ; lower,
and place the round-hole over the square-hole in the trestle-
trees, keeping the bolts in the cap *under*. The topmast is then
hove well through, the men in the top being ready to place
the cap over the head, and lashing it in a secure manner ; a
capstan bar is thrust in the fid-hole with a hauling-line on the
end ; it is then hove high enough for the cap to enter over the
lower mast-head ; haul on the line from the bar in the heel of
the topmast, and it will slue the mast and bring the square hole
of the cap over the lower mast-head ; it is then lowered a
little, and the cap placed, the lashing taken off, then beaten down
into its place.

The cap being fixed securely over the lower mast-head, the
topmast is hung by the up and down tackles, to unreeve the
hawser. The top blocks are unlashed, then hooked to their
proper bolts on each side of the cap, the top-tackle-pendant is
then rove through one block, through the trestle-trees, through
the sheave in the top-mast, up through the trestle-trees again,
and reeve the end through the foremost bolt in the cap of the
opposite side of the block (before reeving it through, parcel
it well) ; take two half hitches on its own, or standing part,
and secure the end with a round seizing. To the lower end
of the top-tackle pendant is hooked (through the thimble) the

block of the top-tackle, connected, by its fall, to a block hooked to an eye-bolt in the deck, and brought to the capstan.

The top-tackle falls and blocks.—The upper block is double, strapped, which is made into a hook; the lower is also double, and should be iron-strapped, having a swivel; a single one is hooked near the double as a leading-block; the fall is rove; the standing part hitched, or clinched, over the block; they are sometimes spliced in, and some have beckets.

To hook the double block, clap a single tail-block well up on the pendant, reeve a whip through it, hitch one end of the whip through one of the sheaves of the double block, hoist it up, and hook it to the pendant.

GETTING TOPMAST CROSS-TREES OVER.

The topmast cross-trees are swayed up in the following manner:—Overhaul a girtline through the round-hole in the cap, and the after girtline outside the top. Hitch the foremost girtline, after it is through the round-hole in the cap, well out on the starboard foremost horns underneath, and secure the end with a seizing of spunyarn; the after one bend on in the same way, to the after-starboard horn; then stop both girtlines well with spunyarn, close to the trestle-trees, and, also, with two stops on the larboard horns. "Sway away," having a guy from the deck to clear it of the top, as it goes aloft. When the upper or larboard horns are well clear of the cap, take two rope's ends from the larboard side of the top, and bend them to the larboard horns, and man them in the top. These are called "steadying lines," and are used to prevent the cross-trees falling back, if a stop is cut *too* soon, and to assist in getting the cross-trees on the cap, and over the mast-head. The cross-trees are swayed higher; and cutting away the stops, and hauling on the steadying lines, the cross-trees will then fall across the cap; place the after hole between the trestle-trees, over the round-hole in the cap—cast off the girtlines and steadying lines—white-lead the mast-head in the wake of the cross-trees, and sway the topmast through; beat the cross-trees well down on the mast-head. The topmast is then swayed a few feet higher, for rigging.

The topmasts are sometimes fidded before rigging, to avoid the greater strain upon the top-tackles.

PLACING TOPMAST RIGGING.

The following sketch exhibits the rigging thus far advanced, and the topmasts struck for placing the topmast rigging, thus:—

Tar the mast-head in the wake of the rigging and clothe the *bolsters* as the lower ones; then place the span for *ginn-blocks*. Some prefer chain spans to shackle the iron-bound block to.

The most approved method is an iron plate with a hook on
each end, which lays across the trestle-trees.   Next put over
the mast-head pendants ; then follow the straps, with thimble
in for standing part of the tyes.   The *shrouds* are swayed and

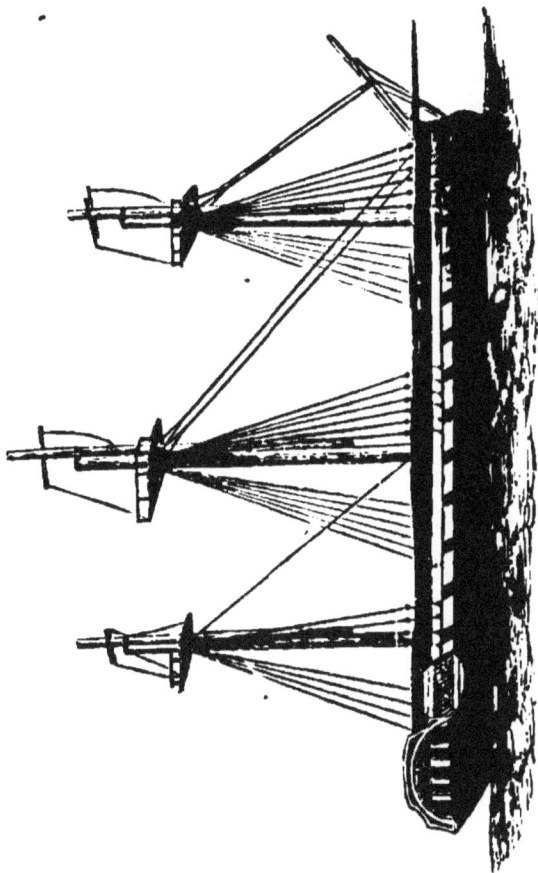

placed over the topmast-head ; the first pair on the starboard
side forward, then the larboard, and so on with the other pairs.
Backstays are hoisted and placed the same as the shrouds ; *stays*
are swayed up and lashed abaft the topmast-head ; the lower
ends reeve through the bees on the bowsprit, and set up to eye-
bolts in the bows with lanyards.

  To seize-in the Sister Blocks.—(See sketch, p. 85.)—They are
seized-in the length of the hanging-block from the eye-seizing, to

prevent any risk of the reef-tackle and lift being jammed between the hanging blocks and the rigging,—one seizing is passed round the shrouds, above the block, another below the block; and a small seizing put on each score round the block and shrouds. The topsail-lift leads through the lower sheave, and reef-tackle through the upper one.

The topmast cap is next swayed up by the girtlines, which are to be lashed well up to the topmast-head for the purpose. Overhaul down before all the foremost ends, and secure them to the foremost bolts in the cap; stop them to the centre ones, and also to the square-hole in the after part; then sway the cap up; when near up, cut the after stops, sway it upon the topmast-head, and the man aloft places it on, then beats it down firmly. The girtlines are unlashed and got down, and the topmast hove up and fiddled.

When the rigging is thus far completed, it is set up in the following manner :—

The puttock-shrouds are set up to the hoop round the mast (see sketch, p. 13), the masts stayed by the burton; the lanyards are rove through the dead-eye in the shrouds, and the dead-eye in the puttock-plate, as the lower rigging, and set up with th . top burton-tackles and runners in large ships.

The standing after-backstays, when in pairs, are fitted with an eye the same as topmast rigging; they are now fitted the same size as lower rigging. The back-stays are set up with a lanyard rove through dead-eyes, the same as shrouds, to a small dead-eye in the after-end of the channel. A service is put on in the wake of the lower yards and tops.

The fore topmast-stays, set up as described, p. 106.

Main topmast-stay is fitted of the same size as the standing back-stay. A large clump block is strapped round the foremast-head, over the eyes of the rigging, and immediately over the square-hole in the after part of the trestle-trees. Through this block the main topmast-stay is rove down through the trestle-trees, it having a thimble turned in the end, a lanyard spliced and rove through it, and set up to a span shackle in the deck, abaft the foremast, for the purpose; or a large bull's-eye hooked to an eye-bolt, and set up on the end.

Main topmast spring-stay leads through a block strapped round the foremast close to the lower rigging, and sets up in the foretop.

Mizen topmast-stay is rove through a thimble strapped round the mainmast-head, over the eyes of the rigging; and when set up, is secured to its own part with round seizings.

When the *shrouds* are again set up for sea, the masts are steadied by their own stays, and not by the burtons.

CHAPTER XII.

— ♦ —

Rigging the Jib-boom—the Traveller, Horses or Foot-ropes, Guys, Martingale-
stay, Martingale-back-ropes, getting the Jib-boom out.—Sending up the
Topgallant-masts—Shrouds, Backstays, the Main Topgallant-stay, the
Mizen Topgallant-stay.—Royal Rigging—the Breast and After back-stay,
Royal-stays, the Fore Royal-stay, the Main Royal-stay, Mizen Royal-stay.
—Ratling the Lower and Topmast Rigging.

HAVING proceeded with the rigging thus far, the nature of its
further progression is such, that many parts may be advancing
at the same time; as, rigging the jib-boom, spritsail-yard, and
whiskers; and getting on board and rigging the fore, main, and
cross jack-yards; then the topsail yards; sending up the top-
gallant-masts, with their rigging and yards, and the flying jib-
boom; rattling the lower and top-mast rigging, &c.

RIGGING THE JIB-BOOM.

The jib-boom being hoisted on board, run the end out on the
bowsprit, pointing it through the stays and bowsprit cap. Reeve
the heel-rope, and sway the jib-boom out a foot or two beyond
the cap. Reeve the jib-stay through the hanks, and hook it to
the *traveller*,—the traveller is first put over the outer end of the
jib-boom, with the hook kept inwards. In some ships the jib-
stay reeves through a sheave-hole, or only a hole, in the boom
end, and a double block turned in the inner end; then a lanyard
or fall is rove through this, and a single block bolted to the bows.
To the traveller seize the jib downhaul blocks and travelling
guys; tar the boom end, put a grommet over, to which seize the
fore topgallant bowline blocks, one on each side (when used).

Horses, or Foot-ropes.—There is one on each side of the jib-
boom, and are fitted thus: take a piece of rope long enough to
make both, cut it in the centre and splice one end into the other
with a cut splice, forming an eye to fit the jib-boom end. Four
or five over-hand knots, or turk's-heads, worked through the
strands, are made at equal distances on the rope from the eye,
for preventing the men from slipping. In each end splice a small
eye, large enough to take a lashing, by which they are set up to
bolts in the bowsprit cap; or the ends are brought in and made
fast, with a round turn round the jib-boom close to the cap.

Guys.—There is one pair on each side; an eye is made to fit the boom end by passing a round seizing when in their place; the inner ends reeve through thimbles on each yard-arm of spritsail yard, or through the sheaves in outriggers, and turn into the strap of a double block, which is connected, by its fall, to a single block, that hooks to an eye-bolt in the bow, or set up to bull's-eyes, and leads upon the forecastle.

Martingale-stay, has an eye in each end to fit the jib-boom, and end of the dolphin-striker. In some ships an iron grommet is fitted with an eye on top and one underneath, neatly leathered, and put over the boom end first; the *martingale-stay* is hooked to the underneath eye, the jib-tack and downhaul to the upper one. Chain martingale is found to answer well in not being liable to stretch.

Martingale back-ropes are pendants, middled and served in the centre, the round of the dolphin-striker, both parts crossed and secured with a throat-seizing, and sets up in board with a tackle. Chain is frequently used in lieu of rope.

Getting the Jib-boom out.—The flying jib-boom iron is driven on after the rigging is placed on the jib-boom; the heel-rope being secured, the boom is hauled out: then the heel-strap is placed in a score in the heel for the purpose, and both bights lashed together; and between the boom and the bowsprit, another lashing is passed round the strap and well frapped together. The heel being secured, the back-ropes and guys are set up.

SENDING UP TOPGALLANT-MASTS.

The topgallant top-blocks being hooked, the *mast-rope* reeves for the topgallant-mast as it does for the topmast: take the end through the square hole in the fore part of the trestle-trees, half-hitch it through the fid-hole, and stop it round the hounds, and the royal mast-head; send the hauling part through the lubber's-hole, and through a leading block or sheave on deck. The topgallant rigging is fitted on a copper funnel, *a*, to slide up and down with the topgallant-mast, which, when struck, rests on the top-mast cap, as the adjoining figures.

Put on the grommet or strap for the main royal stay (if this be the fore topgallant-mast) to reeve through, then put on the *topgallant* and *flying jib-stays*. *Shrouds* next, the same as the topmast. *Breast* and *standing backstays*, the same as the topmast; then the royal-rigging and truck, and reeve the signal halyards.

The *topgallant-mast* is then swayed up and fidded.

Shrouds are set up thus: the ends reeve through the horns of the cross-trees, and set up to an iron *spider hoop* (round the hounds of the topmast) with eyes for the topgallant rigging to lead through.

Backstays set up the same as the topmast backstays.

The fore topgallant-stay reeves through the outer sheave-hole in the jib-boom, then through a sheave-hole in the dolphin striker, and through a bull's-eye hooked to the bows, and when set up, is seized to its own part.

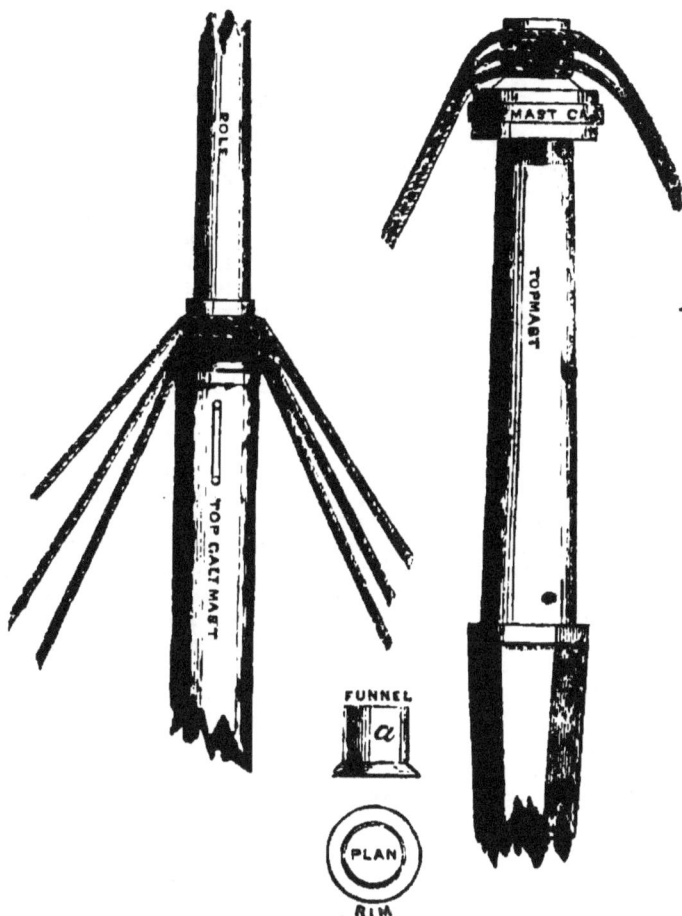

The main topgallant-stay is rove through a block strapped around the fore top-mast head, or through the middle sheave in the after chock of the fore topmast cross-trees, and set up in the fore top.

The mizen topgallant-stay is rove through a bull's-eye in the after part of the main cap, and set up in the main top.

ROYAL RIGGING.

The breast and after back-stay, on each side, are seized as the after back-stays on topgallant-masts. The breast back-stay, or *shroud,* is pulled up with a gun-tackle purchase; the after leg has a thimble turned in, and sets up in the after part of the channels, with a lanyard. The *shrouds* are set up in the top (breast back-stay fashion).

Royal-stay.—Splice an eye in the stay to fit the mast-head, cover it, and serve over the splice. It goes on next to the grommet, then the shroud and back-stays, spanned together.

The fore-royal-stay is rove through the outer sheave-hole in the flying-jib-boom end, and through a hole in the dolphin-striker, or pulled up through a fair leader on the forecastle.

The main-royal-stay is rove through a thimble stopped around the fore-topgallant-masthead, through another strapped round the eye of a shroud, and when set up is seized to its own part.

Mizen-royal-stay reeves through a sheave in the after part of the main-topmast trestle-trees, or through a block strapped around the main-topmast-head, through a thimble strapped round the eye of a main-shroud, and seized to its own part.

RATLING THE LOWER AND TOPMAST RIGGING.

The puttock-shrouds are set up to the hoop around the mast (see p. 13); topmast stayed, rigging and back-stays set up, lanyards secured as lower rigging. Small spars as boat's oars, or anything *light* that will answer, are seized to the shrouds, about four or five feet asunder, for the men to stand upon whilst ratling. The ratlings are fastened round each shroud with a clove-hitch, except at the ends, small eyes are spliced in, and seized to the shroud: in three or four places take a ratling to the after swifter—these are called *shear ratling.*

The ratlings are fastened horizontally to the shrouds, at distances of thirteen to fifteen inches from each other. Every man employed should have a *measure* within his reach, and care should be taken to make the ratlings on one side correspond in a parallel direction with those of the other.

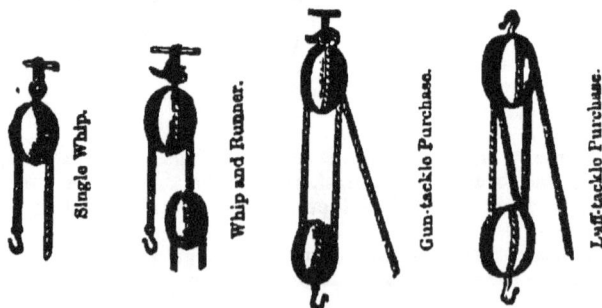

Single Whip. Whip and Runner. Gun-tackle Purchase. Luff-tackle Purchase.

CHAPTER XIII.

—✦—

Rigging the Fore and Main-yards.—Trusses to Lower-yards, Iron Jack-stays, Quarter or Topsail-sheet Blocks, Clue Garnet Blocks, Leech-line Blocks, Bunt-line Blocks, Lift Blocks, Foot-ropes and Stirrups, Brace Blocks, Fore Braces, Main Braces.—Rigging the Topsail-yards—Iron Jack-stays, Earing Strap, Foot-ropes and Flemish Horses, Brace Blocks, Fore and Main Top-sail Braces, Mizen Top-sail Braces, Lift Blocks, Tye Blocks, Quarter Blocks, Topsail Tyes, the Fly Blocks, Reeving Topsail Halliards, the Mizen Topsail Halliards, Bunt-line Lizards.—Rigging the Topgallant Yards, the Tye or Halliards, Fore Topgallant Braces, Main Topgallant Braces, Mizen Topgallant Braces, Clue-line and Royal-sheet or Quarter Blocks.—Rigging the Royal Yards—Rigging the Mizen or Spanker Gaff, Reeving Throat Halliards, Reeving Peak Halliards, Fitting Checks or Brail Blocks, to fit Single Vangs.—Rigging the Spanker Boom, the Topping Lifts, Spanker-boom, Sheet and Guys in one.—A Brig or Schooner's Main-boom.

RIGGING THE FORE AND MAIN YARDS.

TRUSSES TO LOWER YARDS.

MERCHANT vessels in general have iron trusses, figs. 1, 2, and 3, for the ease of bracing the yards, y, up. The mechanism may be described as follows:—h is the hoop on the mast; ff, the hoops on the yard; $a\,a$, the universal joint; s, the screw for setting the hoop tight on the mast; and c, for fixing it to the yard.

Fig. 1.

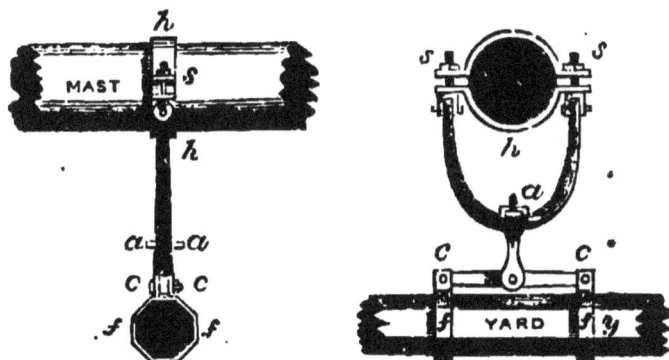

Fig. 2.

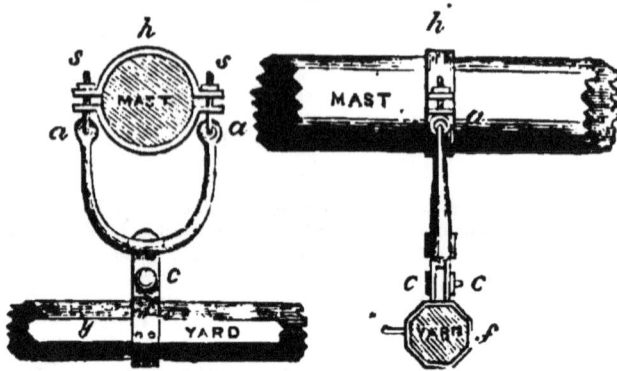

Fig. 3.

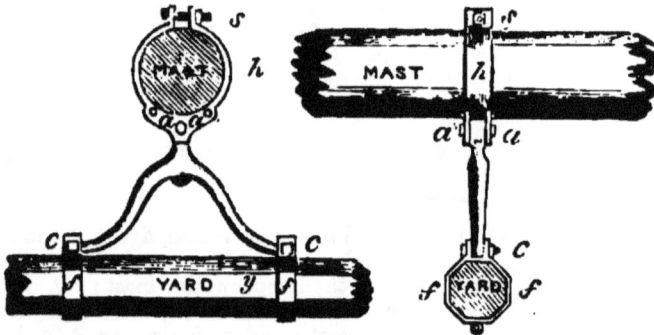

Jack-stays.—Iron jack-stays for yards are used in all merchant vessels; they reeve through small eye-bolts, driven into or eyes in hoops round the yard, one on each side of the middle or slings of the yard, and are for the purpose of bending the sails to. The cross jack-yard has no jack-stay.

Quarter or topsail-sheet blocks are iron blocks fitted to the quarter-hoops on each side of the middle or slings of the yard ; or fitted each to the iron hoop in the slings of the yard for that purpose ; the chain topsail-sheets reeve on their respective sides, and lead down by the mast.

Clue-garnet blocks are iron-bound blocks fitted to the quarter-hoops, when the topsail-sheet blocks are fitted to the sling-hoop.

Leech-line blocks are seized to the iron jack-stay on the fore-part of the yard, one-fourth within the cleets on each yard-arm. There are two leech-lines in large ships.

Bunt-line blocks are hooked to eye-bolts underneath the top between the forepart of the trestle-trees.

Lift-blocks are single; an iron plate is bolted across the upper side of the main or fore-cap; it is in the form of a crescent, with the hollow side towards the top-mast. In each end of the crescent, or horn, an eye or hook is turned, and the blocks attached to each. (See sketch, p. 26.) The lifts go over the yard-arms, with an eye spliced in the end to fit them. The other is rove through the block at the cap, and through lubber's-hole on deck, where they are belayed.

Foot-ropes and Stirrups.—The foot-ropes are cut once-and-a-half the length of the yard, excepting lower-yards. An eye, to fit the yard-arm, is spliced in their outer ends, and hang about three feet below the yards. To keep the foot-ropes more parallel to the yards, it is suspended at proper distances, by short pieces of ropes spliced round the foot-rope, called *stirrups;* sometimes two, three, or four on each side of the yard: eyes spliced in the opposite ends, or seized to the jack-stays. The inner ends of the foot-ropes have a small eye, to take a seizing to the jack-stays and round the yard, next the slings.

Brace-blocks are next put over the yard-arms; some go with rope or chain pendants. The block is a large single one, with two scores for rope, and iron-bound for chain, through which the brace reeves.

Fore-braces are clove-hitched, and the end seized aft on the collar of the main-stay, below the splice; the other end taken forward and rove from in (out), through the block on the yard, through a single block strapped into a bolt in the cheek of the main-mast, with a thimble in it, close up to the trestle-trees, then rove from forward aft, through a sheave in the main fife-rail.

The *brace* is often middled, and clove-hitched in the bight on the main-stay, and both ends taken forward and rove as before. (See plate, p. 92.)

Main-braces.—The standing part of the brace being parcelled, is clenched round a bumkin, or an iron out-rigger on the quarters, for the purpose; the hauling-part reeves through the block at the yard-arm, and back through a block which is strapped to the bumkin end; then through a sheave in the bulwarks (abaft), for the purpose, and belays round a cleat inside.

The blocks are on the fore-side of the main-yard in *brigs,* and the brace reeves through a block strapped into a bolt, with a thimble in it, in the sides of the fore-mast trestle-trees, then rove from aft forward, through a truck seized on to the after-most fore-mast-shrouds, and belays as before. The braces are rove when the yards go up into their respective places.

The yards being rigged are sent aloft as follows:—The end of the hawser is rove through the block at the lower mast-head, and overhauled down, and made fast to the slings of the yard; then securely stopped along the yard in several places, and also at the upper yard-arm. As it comes on board, the stops

are cut, and easing away on the pendant tackle, then bousing on
the other, as the yard advances on board beyond the slings.
The yards are placed square, before their respective masts; the
hawser is hove upon until the yard is high enough to shackle
the chain—slings which are put round the mast-head to hang
the yard by: square the yard by braces and lifts, and cast off
and unreeve the hawser; then secure the iron trusses on the
yard to the mast.

RIGGING THE TOP-SAIL-YARDS.

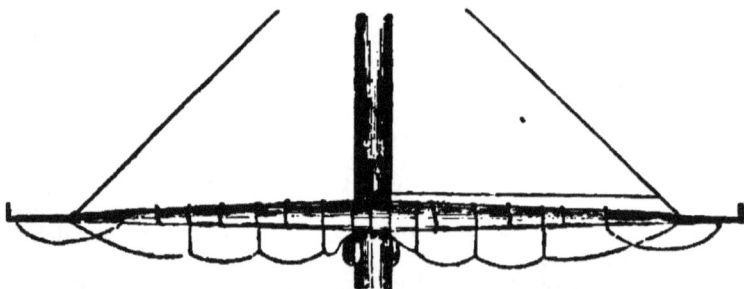

The iron jack-stays are rove through the eye-bolts or staples,
driven into the yard, and forelocked the ends a-midships: the
rigging is then placed on the yards as follows:—First, the
earing-strap with a small thimble seized-in; the *foot-ropes* next,
the same as the lower yards, with the addition of *Flemish horses*,
which have an eye spliced in each end; one eye is spliced
round a thimble, which is on the neck of the pacific-iron, or
boom-iron, on the yard-arm, and the other is seized round the
yard within the arm-cleats.

Brace-blocks are strapped in the same way as the fore or main-
yard. *The foretop-sail braces* reeve through the block at the
yard-arm, and then taken to the maintop-mast head, where it
is rove through a block lashed on each side for the purpose,
from thence on deck; the standing-part makes fast to the
collar of the main-stay. Brigs the same.

Main-topsail braces reeve the same as the main braces; the
standing-part makes fast to the collar of the mizen-stay. Brigs;
—the standing-part makes fast to the after-end of the fore-mast
cap. The leading part leads forward through a single block,
seized into a single strap, and secured to an eye-bolt on each
side of the fore-cap, and through the lubber's-hole on deck.

Mizen top-sail braces reeve through the block at the yard-arm,
and cross as the cross-jack braces; but the lead is at the mainmast-head, instead of the shrouds.

Lift-blocks, in large ships, are strapped with an eye to the
size of the yard-arm. The lifts reeve through the lower sheave

in the *sister block* in the top-mast shrouds, and through the block in the yard-arm. The standing-part is secured round the top-mast head, and the leading-part leads down the side of the mast, and sets up in the top, or on deck. The lifts are single in the merchant-service.

Tye-blocks are now generally iron-bound, swivel-fashion, and bolted into an eye in the hoop round the yard for the purpose.

Quarter-blocks are double blocks iron-bound, and secured in the same way as tye-blocks ; through which the clue-line and topgallant-sheet reeves, and leads down upon deck.

Top-sail tyes.—Large ships have double tyes ; the lower end has a single or lower block, called the *Fly-block*, spliced for the halliards. The upper-end is first rove through the bullock-block from aft, then through the tye-block on the yard, and the end taken to the mast-head, so that it can be shortened up.

The fly-blocks are large flat blocks ; some double, sometimes single, and often one double and one single to each.

Reeving Top-sail Halliards.—When rove double, a single block is strapped into, or hooked, to a swivel-bolt in the after-part of the chains ; one end of the halliards is spliced into the upper part of the strap of this block, or bent into a becket put there for the purpose, and the end seized. The other end is then rove through one of the sheaves in the double block in the tye, then through the sheave in the single block in the chains, through the other sheave in the double block, and through a leading block on deck.

The mizen top-sail halliards have only one tye. The standing part is clenched or half-hitched to the strap, with the thimble, at the mizen top-mast head, and a single block spliced or secured in the end. Another single block is strapped into a swivel-bolt in the mizen chains, and the halliards rove as with two single blocks ; the fall rove through a leading block or cheek.

Bunt-line lizards are spliced round the strap of the top-sail tye-block upon the yard.

The yard is next hove up, and the *parral* is passed round the aft part of the mast and seized to the rolling cleats, or jaws fixed on the aft side, the middle, or slings of the yard. The patent iron parrals are now commonly used in merchant ships.

RIGGING THE TOPGALLANT-YARDS.

They are got on board like the topsail-yards. First, leather the rolling-cleats, seize on the parral and quarter-blocks,—foot-ropes the same as topsail-yards — stirrups one to each foot-rope—iron jack-stays secured to the yard with staples, fitted the same way as topsail-yards. Iron sling-hoops as other yards.

The lifts are single ; an eye is placed to fit the yard-arm ; the other end is rove through the thimble, or bull's-eye in the topgallant-shrouds ; a thimble turned into the end, and a lanyard spliced into it, and set up to another thimble strapped round a puttock-plate inside the dead-eye in the top, or set up on the end.

The tye or halliards reeves through the sheave-hole in the head of the topgallant-mast, and clinches or shackles to the eye in the sling-hoop; the lower end of the halliards comes down abaft the mast, upon which any required purchase is added.

Fore Topgallant-braces.—The standing-part makes fast round the yard-arm, and the leading part reeves through a single tail-block, secured to the first and second shrouds of the main top-mast rigging, through lubber's-hole, and through a fair leading sheave on deck.

Main topgallant-braces are single, and go with an eye over the yard-arm. They lead aft to the mizen topmast-rigging, and are fitted to reeve the same as the fore; the hauling part before all, through lubber's-hole, and through a sheave in the rack, or a leading block, to the side abreast of the mizen-mast.

Mizen topgallant-braces are single. An eye is spliced in one end to fit the yard-arm. The other end is rove through a single block, secured to an eye-bolt on each side of the main cap, and through lubber's-hole on deck.

Clue-line and royal sheet, or quarter-blocks, are double, which are strapped with two lashing-eyes, and lash together on the top of the yard—the foremost sheave for topgallant clue-line, after one for royal sheet. The leading part leads down the mast on deck.

The yard is swayed up, and the parral fixed.

RIGGING THE ROYAL YARDS.

They are fitted the same as topgallant-yards, with little exceptions.

RIGGING THE MIZEN OR SPANKER GAFF.

Some ships in the merchant service have their gaffs fitted to hook to an iron hoop, with a hook or goose-neck in the end of the gaff, instead of jaws (see p. 32). Others travel up and down an iron groove or railway, fitted to the lower mast (using no trysail-mast). Others again use a wooden batten nailed to the mast; some an iron jack-stay, and some a rope one. The trysail-masts are preferable in a *gale of wind*.

Reeving Throat-halliards.—The standing part is spliced into the single-block, which is hooked to the gaff, up through the double block under the top, down through the block, up again through the other sheave in double block, and through a leader, opposite to the peak-halliards.

Reeving Peak-halliards.—The end of the halliard is taken through the lubber's-hole, rove through a double-block at the mizen-cap, which block hooks to an iron strap over the cap, down through the inside block, d (see sketch, p. 32), on the gaff, up again through the double-block before described, then through a block further out, d, on the gaff; and the standing part is either hitched round the head of the mizen-topmast, or made fast round the neck of the block, at the after-part of the cap. The standing-part may be spliced round the gaff, by dispensing with one block.

Fitting Cheeks or Brail-blocks.—Gaffs are mostly fitted with cheeks, instead of blocks; and sheaves cut in the jaws for the throat-brails, and fair leaders; which is the approved plan at present, and is very neat.

To fit single Vangs.—Middle the required length of rope, and seize the bights to fit the gaff-end, and lead one end on each side. Vangs steady the gaffs amidships.

All gaffs should be peaked, or elevated to an angle parallel with the mizen-topmast stay.

RIGGING THE SPANKER-BOOM.

The topping lifts have hooks spliced in the end, which hook to eyes in one, and sometimes two, hoops, d, e, round the boom (see sketch, p. 32); the ends are rove through a single block strapped into a bolt with a thimble in it on each side of the mizen-trestle-trees; and splice a parcelled thimble in the end, for the purpose of hooking the jigger-tackle.

Spanker-boom-sheet and Guys in one.—The boom-sheet is fitted thus;—into a bolt, with a thimble in each quarter, strap a double block with a single strap; then seize into two grommet-straps, worked round the boom (wormed and covered), two single blocks, one on each side, just the outside of the taffrail. Secure these blocks in their straps with a round seizing passed between the block and the boom. The rope for the guys is middle and cut; then an eye, or cut splice made to fit the boom-end. Take the larboard-guy, and reeve it through one of the sheaves in the double block on the quarter, through the single block on the boom, through the other sheave in the double block, through a fair leader in the side, and pull it upon deck. The starboard one is rove in the same manner through the block on the starboard quarter.

A BRIG OR SCHOONER'S MAIN-BOOM.

These booms having so little projection over the stern, guys are unnecessary. On each quarter strap a double block, and one on each side of the boom, in separate straps: through these reeve the sheet, the standing part from the strap of the quarter-block, and hauling part through one of the sheaves of the quarter-block.

CHAPTER XIV.

Rigging a Brig.— Brigantine.— Schooner.— Steamer.— Table showing the comparative strength of Chain Rigging and Hemp Rigging.—Tables of the size of the Standing and Running Rigging of Ships; with the description, size, and number of blocks, hearts, dead-eyes, &c.—Tables of the size of the rigging for schooners.—Tables of the size of the rigging for Cutter-Yachts.—A Table showing the comparative sizes, weight, and strength between Newall and Co's Patent Wire Ropes and Hemp Ropes, for standing Rigging.—A Table showing the comparative strength between Iron Chains and Hemp Ropes.—A Table showing the strength of short round-linked Bobstay, Bowsprit-shroud or Crane Chain, without studs, such as is used for Rigging, &c.—A Table showing the weight of Chain Cable.—Cordage Table, showing the weight of one Fathom of Rope, from 1 to 24 inches inclusive, plain-laid 3 strand, such as used for running rigging, &c.—A Table of the weigh. of Tarred Cordage.—A Table showing the weight of 100 Fathoms of Cable-laid Rope, from 1 to 24 inches. Also a comparative size of Chain.—A Table showing the strength of Plain-laid Rope of three strands.

RIGGING A BRIG.

THE rigging of a brig is so little different from that of the fore and main-masts of a ship, that the description of the one will serve equally well for the other. It may be observed, however, that the braces of the yards on the main-mast lead forward, and are sometimes small chain; the lifts of top-sail and topgallant-yards are fixtures to the mast-heads. In brigs as well as ships a great quantity of mats is used in the way of chafes against the rigging; such as the foremost swifters of the lower rigging and back-stays, on account of the foot and clew of the courses, when reefed and hauled aft, grinding against them high up. To take the chafe off the foremast shrouds of the topmast rigging, when the topsail yards are braced up, a quarter mat abaft the yards on each side is required. For the back-stays, in the wake of the lower yards, when braced up, mats or platting, or some other substitute, is necessary as a protection. Merchant vessels have these places served and use *scotsmen* (slips of wood so named); but leather neatly stitched on is the best.

RIGGING A BRIGANTINE.

A *Brigantine* is a vessel rigged the same way as a brig on the foremast, and similar to a schooner on the mainmast.

RIGGING A SCHOONER.

A *Schooner* is a vessel with two masts and a bowsprit, whose mainsail and foresail are both suspended by gaffs, like a sloop's mainsail; the masts rake aft, but the bowsprit lies nearly horizontal; also a jib-boom and topmasts. The main-stay leads to the head of the foremast; also two jumper-stays, which set up to an eye-bolt in the deck, just at the after-part of the fore-rigging, so that the weather one is always kept taut or tight.

RIGGING A STEAMER.

A Steamer has one to three masts and a bowsprit; the fore-mast has a topmast and topgallant-mast, on which is set

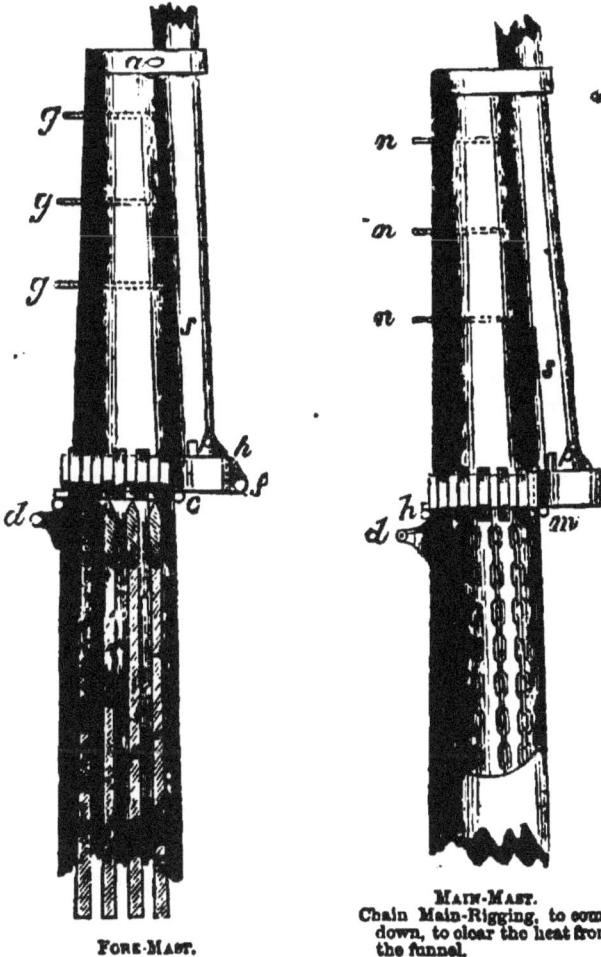

FORE-MAST.

MAIN-MAST.
Chain Main-Rigging, to come down, to clear the heat from the funnel.

a fore-boom-sail, top-sail, and top-gallant-sail, similar to a schooner's; abaft the main and mizen-masts are a boom-main-sail and mizen. The main and mizen-masts have topmasts, and occasionally carry gaff-top-sails. The bowsprit is short, nearly horizontal, with a jib and flying jibboom in one, upon which the head sails are set.

In the annexed sketches the rigging attached to the iron band round each of the mast-heads is shown thus:—*The rope shrouds* have iron sockets riveted through the ends of them, having eyes formed in the ends of the sockets, for connecting the shrouds to links fitted over the iron-band round the mast-head. They are locked together by a bolt passing through the links and eyes between them, thus completely securing the rigging to the mast-head in a very snug manner.

Steam vessels have a great quantity of chain about the main-mast head on account of the heat from the funnel; such as chain-shrouds about one-third of their length, down from the mast-head, which are similarly secured as the fore-mast shroud, the most part of the main-stay is chain. The peak-halliards, topsail braces, maintop-mast stay, throat-halliards, boom-topping lifts, &c., consists of chain; for which purpose, eye-bolts are fitted in the iron bands round the masts; as, f for slinging the fore-yard; h, for shackling a block for jib-halliards: c, for staysail-halliards; d, for throat-halliards; h, (on the main-mast) for boom-topping-lift; m, for main-stay; a, in the iron-cap at the fore-mast head, for hooking a block for the lifts of the fore-yard: $g\,g\,g$ and $n\,n\,n$, for peak-halliards; s, a sheave in the heel of the top-mast, and p, the fid-hole.

TABLE SHOWING THE COMPARATIVE STRENGTH OF CHAIN-RIGGING, SUCH AS IS GENERALLY USED IN STEAM VESSELS AND HEMP-RIGGING.

Chain Rigging.	Equal to Hemp Rigging.	Weight of Chain per Fathom.	Chain Rigging.	Equal to Hemp Rigging.	Weight of Chain per Fathom.
$\frac{5}{16}$ in.	$1\frac{3}{4}$ in.	4 lb.	$\frac{7}{16}$ in.	$4\frac{1}{2}$ in.	20 lb.
$\frac{3}{8}$,,	2 or $2\frac{1}{4}$,,	6 ,,	$\frac{5}{8}$,,	5 or $5\frac{1}{4}$,,	25 ,,
$\frac{7}{16}$,,	$2\frac{1}{4}$,, $2\frac{3}{4}$,,	7 ,,	$\frac{11}{16}$,,	6 ,, $6\frac{1}{2}$,,	29 ,,
$\frac{1}{2}$,,	3 ,, $3\frac{1}{4}$,,	10 ,,	$\frac{3}{4}$,,	7 ,, $7\frac{1}{2}$,,	36 ,,
$\frac{9}{16}$,,	$3\frac{3}{4}$,,	12 ,,	$\frac{7}{8}$,,	$6\frac{1}{2}$,, 9 ,,	48 ,,
$\frac{5}{8}$,,	4 ,,	17 ,,	1 ,,	10 ,,	63 ,,

Recently WIRE ROPES, in lieu of chain, have been used for standing rigging, and they have been found to answer the purpose very well for making shrouds, and for all standing stays, as in such cases they are only applicable; and when once set properly tight, they do not run up when wet, or stretch like hempen ropes. See table showing the comparison between wire and hempen ropes, p. 140.

TABLE SHOWING THE SIZE OF STANDING AND RUNNING RIGGING OF MERCHANT SHIPS.

Names of Rigging.	Ship of 1100 tons.				Ship of 800 tons.				Ship of 600 tons.				Barque of 450 tons.				Barque of 300 tons.				Brig of 200 tons.			
	Size of Rope in Inches.	Blocks, &c.			Size of Rope in Inches.	Blocks, &c.			Size of Rope in Inches.	Blocks, &c.			Size of Rope in Inches.	Blocks, &c.			Size of Rope in Inches.	Blocks, &c.			Size of Rope in Inches.	Blocks, &c.		
		Description.	Size in Ins.	Number.		Description.	Size in Ins.	Number.		Description.	Size in Ins.	Number.		Description.	Size in Ins.	Number.		Description.	Size in Ins.	Number.		Description.	Size in Ins.	Number.
Bowsprit Gear.																								
Gammoning (chain for all classes of vessels)	¾	..	..	..	1 1/16	..	..	..	⅝	..	..	..	⅝	..	..	..	9/16	..	..	..	½	..	..	..
Shrouds (chain)	1 1/16	H	10	4	1 1/16	H	9	2	⅝	H	8	2	⅝	H	7	2	9/16	H	6	2	9/16	T	..	2
Laniards for shrouds (four-stranded	3	..	..	..	2½	..	..	..	2½	..	..	..	2½	..	..	..	2	..	..	..	2	..	..	..
Bobstays (chain)	⅞	H	12	3	¾	H	10	3	¾	H	9	2	1 1/16	H	8	2	⅝	H	7	2	⅝	T	..	2
Laniards for bobstays	4	..	..	..	4	..	..	..	3½	..	..	..	3	..	..	..	2½	..	..	..	2	..	..	..
Man-ropes	4	T	..	4	3½	T	..	4	3½	T	..	4	3	T	..	4	3	T	..	4	2½	T	..	4
Jib-Boom Gear.																								
Jib-stay and strapping	5½	C	11	1	5	C	10	1	4½	C	9	1	4	C	8	1	4	C	8	1	3½	C	8	1
Guys, single	5	..	..	..	4½	..	..	..	4½	..	..	..	4	..	..	..	4	..	..	..	4	..	..	..
Falls	3	D	9	4	2½	D	8	4	2½	D	8	4	2	D	7	4	2	D	7	4	2	D	..	4
Foot-ropes	3	..	..	..	3	..	..	..	2½	..	..	..	2½	..	..	..	2½	..	..	..	2½	..	..	..
Martingale stay	7	..	..	..	6½	..	..	..	6	..	..	..	6	..	..	..	5½	..	..	..	4½	..	..	..
Martingale backropes	4½	..	..	..	4	..	..	..	4	..	..	..	3½	..	..	..	3	..	..	..	2½	..	..	..
Martingale falls	2½	D / S*	8 / 8	2 / 2	2	D / S*	7 / 7	2 / 2	2	D / S*	7 / 7	2 / 2	1¾	D / S*	6 / 6	2 / 2	1½	D / S*	6 / 6	2 / 2	1½	T	..	2
Halliards	3½	S	10	1	3	S	9	1	3	S	9	1	3	S	9	1	2½	S	8	1	2	S	7	1
Downhaul	2½	S	8	1	2¼	S	8	1	2	S	7	1	2	S	7	1	2	S	7	1	2	S	7	1
Sheets	3	S	9	2	2½	S	8	2	2½	S	8	2	2¼	S	8	2	2¼	S	8	2	2	S	7	2
Pendents	4½	..	..	..	4½	..	..	..	4	..	..	..	4	..	..	..	3½	..	..	..	3	..	..	..

FLYING-JIBBOOM GEAR					
Flying-jib stay					
Guys					
Stay tackle falls					
Foot-ropes					
Martingale stay					
Halliards and stripping					
Down-haul and stripping					
Sheets					
Heel-lashing					
FORE AND MAIN-MASTS.					
Pendants					
Shrouds					
Lanyards for shrouds					
Stays					
Collars for stays					
Runners of tackles*					
Falls of tackles					
Ratlines					
FORE AND MAIN-YARDS.					
Slings proper (to go over cap), chain for all vessels					
Jackstay (bending) iron					
Foot-ropes					
Stirrups					
Lifts					
Braces (4 in number)					
Earings (tapered) or chain					
Sheets (tapered)					
Clue-garnets					

* Double-blocks: the upper one is lashed to the pendant.

TABLE SHOWING THE SIZE OF STANDING AND RUNNING RIGGING OF MERCHANT SHIPS.

Names of Rigging.	Ship of 1100 tons. Size of Rope in inches.	Ship of 1100 tons. Blocks &c. Description.	Ship of 1100 tons. Blocks &c. Size in Ins.	Ship of 1100 tons. Blocks &c. Number.	Ship of 800 tons. Size of Rope in inches.	Ship of 800 tons. Blocks &c. Description.	Ship of 800 tons. Blocks &c. Size in Ins.	Ship of 800 tons. Blocks &c. Number.	Ship of 600 tons. Size of Rope in inches.	Ship of 600 tons. Blocks &c. Description.	Ship of 600 tons. Blocks &c. Size in Ins.	Ship of 600 tons. Blocks &c. Number.	Barque of 400 tons. Size of Rope in inches.	Barque of 400 tons. Blocks &c. Description.	Barque of 400 tons. Blocks &c. Size in Ins.	Barque of 400 tons. Blocks &c. Number.	Barque of 300 tons. Size of Rope in inches.	Barque of 300 tons. Blocks &c. Description.	Barque of 300 tons. Blocks &c. Size in Ins.	Barque of 300 tons. Blocks &c. Number.	Brig of 200 tons. Size of Rope in inches.	Brig of 200 tons. Blocks &c. Description.	Brig of 200 tons. Blocks &c. Size in Ins.	Brig of 200 tons. Blocks &c. Number.
FORE AND MAIN-YARDS—*continued.*																								
Bowlines and bridles																								
Buntlines and falls																								
Leech-lines																								
Slab-line and strapping																								
Jigger-falls and strapping																								
Fore-stay-sail-stay																								
Halliards																								
Sheets																								
Tack-lashing																								
Downhaul																								
Lower studding-sail—Halliards																								
Inner halliards																								
Span for outer halliard																								
Lower studding-sail—Sheets																								
Tack																								
Tripping-line																								
Strapping and tailling																								
Swinging-boom guys																								

FORE AND MAIN-TOP-MAST.

Shrouds				D E		
Lanyards for shrouds						
Ratlines						
Standing backstays				D E		
Lanyards for backstays						
Burton pendants				T		
Falls and strapping				PL.DE		
Stays						
Lanyards						
Futtock shrouds						
Lashings						
Ratlines						
Stay-sail halliards						
Downhaul						
Strapping						
Pendants						
Sheets						
Tack-lashing						

FORE AND MAIN-TOP-SAIL YARDS.

Top-sail tyes (all chain)				S.I.bd		St.bd
Halliards for top-sail tyes				D		
Strapping bullock blocks				S*		
Jackstays (iron)				S		
Foot-ropes				T		
Stirrups				S		
Flemish horses				S		
Braces				Sin.		
Lifts						
Parral-rope						
Clue-lines and strapping				St.bd		

TABLE SHOWING THE SIZE OF STANDING AND RUNNING RIGGING OF MERCHANT SHIPS.

Names of Rigging.	Ship of 1100 tons.				Ship of 800 tons.				Ship of 600 tons.				Barque of 400 tons.				Barque of 300 tons.				Brig of 200 tons.			
	Size of Rope in Inches.	Blocks, &c.			Size of Rope in Inches.	Blocks, &c.			Size of Rope in Inches.	Blocks, &c.			Size of Rope in Inches.	Blocks, &c.			Size of Rope in Inches.	Blocks, &c.			Size of Rope in Inches.	Blocks, &c.		
		Description.	Size in Ins.	Number.		Description.	Size in Ins.	Number.		Description.	Size in Ins.	Number.		Description.	Size in Ins.	Number.		Description.	Size in Ins.	Number.		Description.	Size in Ins.	Number.
FORE AND MAIN-TOP-SAIL-YARDS—*continued.*																								
Bunt-lines and strapping	3	S	9	2	3	S	9	2	2½	S	8	2	2½	S	8	2	2½	S	7	2	2	S	7	2
Span	3	T	..	2	3	T	..	2	2½	T	..	2	2½	T	..	2	2	T	..	2	2	T	..	2
Bow-lines and strapping	3	S	9	2	3	S	9	2	3	S	9	2	2½	S	8	2	2½	S	8	2	2	S	7	2
Reef-tackles and strapping	3	S	9	2	3	S	9	2	3	S	9	2	2½	S	8	2	2½	S	8	2	2	S	7	2
		T		2		T		2		T		2		T		2		T		2		T		2
Sheets (all chain)	..	..	..	..	..	..	..	..	..	..	..	..	..	..	..	..	..	..	..	..	..	..	..	..
Studding-sail halliards	3½	S	11	4	3½	S	11	4	3½	S	11	4	3	S	10	4	2½	S	10	4	2	S	9	4
Sheets	3¼	..	..	..	3	..	..	..	3	..	..	..	3	..	..	..	2½	..	..	..	2	..	..	..
Tacks	3½	S	11	4	3½	S	11	4	3½	S	11	4	3	S	10	4	2½	S	10	4	2	S	9	4
Downhaul	2	S	6	2	2	S	6	2	1½	S	5	2	1½	S	5	2	1½	S	5	2	1½	S	5	2
Boom-jiggers	2	D	7	2	2	D	7	2	2	D	7	2	2	D	7	2	..	..	..	..	..	..	..	..
		S	7	2		S	7	2		S	7	2		S	7	2								
Heel-lashing	3	..	..	..	3	..	..	..	2½	..	..	..	2½	..	..	..	2	..	..	..	2	..	..	..
Boom-brace-pendant	3½	S	6	2	3	S	6	2	3	S	6	2	3	S	6	2	2½	S	6	2	2½	S	6	2
Whip	2	..	..	..	2	..	..	..	2	..	..	..	2	..	..	..	2	..	..	..	2	..	..	..
FORE AND MAIN-TOP-GALLANT-MASTS.																								
Shrouds	4	T	..	8	3½	T	..	8	3½	T	..	8	3	T	..	8	3	T	..	8	2½	T	..	8
Laniards for shrouds	2	..	..	..	1½	..	..	..	1½	..	..	..	1¼	..	..	..	1	..	..	..	1	..	..	..
Backstays	4½	D E	6	4	4	D E	6	4	3½	D E	6	4	3	T	..	4	3	T	..	4	2½	T	..	4
Laniards	2	..	..	..	2	..	..	..	1½	..	..	..	1½	..	..	..	1½	..	..	..	1½	..	..	..

Stay
Tackle-fall and strapping
Royal-stay
Backstays
Laniards

FORE AND MAIN-TOP-GALLANT-YARDS.
Halliards and strapping
Jackstay (iron)
Foot-ropes
Braces and strapping
Lifts
Parral-ropes
Clue-lines
Straps for quarter-blocks
Bow-lines and strapping
Bridles
Sheets
Earing
Studding-sail-halliards
Sheets
Tacks and strapping
Downhauler and straps.

FORE AND MAIN ROYAL YARDS.
Halliards
Jackstay (iron)
Foot-ropes
Braces and strapping
Lifts
Parral-lashing
Clue-lines and strapping
Bow-lines
Sheets
Earings

TABLE SHOWING THE SIZE OF STANDING AND RUNNING RIGGING OF MERCHANT SHIPS.

Names of Rigging	Ship of 1100 tons — Size of Rope in Inches	Blocks &c. Description	Size in Ins.	Number	Ship of 900 tons — Size of Rope in Inches	Description	Size in Ins.	Number	Ship of 600 tons — Size of Rope in Inches	Description	Size in Ins.	Number	Barque of 400 tons — Size of Rope in Inches	Description	Size in Ins.	Number	Barque of 300 tons — Size of Rope in Inches	Description	Size in Ins.	Number	Brig of 200 tons — Size of Rope in Inches	Description	Size in Ins.	Number
MIZZEN-MAST.																								
Shrouds	6½	D E	9	12	6	D E	8	10	6	D E	8	8	5½	D E	8	8	5	D E	8	8	:	:	:	:
Lanïards for shrouds	3			:	4			:	2¾			:	2¾			:	2¾			:	:	:	:	:
Burton-pendants	4½	T	10	2	2½	T	8	2	2	T	7	2	3	T	7	2	2	T	7	2	:	:	:	:
Falls and strapping	3	D* S*	10	2	2	D* S*	8	2	2	D* S*	7	2	2	D* S*	7	2	2	D* S*	7	2	:	:	:	:
Ratlines	1¼	T		1	1½	T		1	1	T		1	1	T		1	1	T		1	:	:	:	:
Stay	7½				7				6½				6				6				:	:	:	:
Seizings	1½				1¼				1¼				1¼				1½				:	:	:	:
Lanïards	2½				2¼				2½				2				2½				:	:	:	:
CROSS-JACK-YARD.																								
Slings (chain for all vessels)	:			:	:			:	:			:	:			:	:			:	:	:	:	:
Trusses (iron)	:			:	3			:	2¼			:	:			:	:			:	:	:	:	:
Foot-ropes	2¼	T		2	2	T		2	2¼			2	:			:	:			:	:	:	:	:
Stirrups	2¼				2¼				2¼				:				:				:	:	:	:
Lifts	2¼				2¼				2¼				:				:				:	:	:	:
Braces and strapping	8	B	8	4	8	B	8	4	8	B			:			4	:			:	:	:	:	:
MIZZEN-TOP-MAST.																								
Shrouds	5	D E	8	8	4½	D E	7	8	4	D E	7	6	3½	D E			3½	D E			:	:	:	:
Lanïards for shrouds	2½				2				2				2				1½				:	:	:	:

Stay		
Backstays		
Breaststays		
Lanyards		
Futtock shrouds		

MIZEN-TOP-SAIL-YARD.

Top-sail ties (chain)		
Halliards for tye and straps		
Jackstay (iron)		
Foot-ropes		
Stirrups		
Flemish horses		
Parral-rope		
Lifts		
Braces		
Sheets (chain)		
Clue-lines and strapping		
Bunt-lines and strapping		
Span		
Bow-lines and strapping		
Bridles		
Reef-tackles		
Earings		

MIZEN-TOPGALLANT-MAST.

Shrouds		
Lanyards		
Backstays		
Lanyards		
Stay		
Lanyard		
Royal-stay		
Backstays		
Lanyards		

TABLE SHOWING THE SIZE OF STANDING AND RUNNING RIGGING OF MERCHANT SHIPS.

Names of Rigging.	Ship of 1100 tons. Size of Rope in Inches.	Ship of 1100 tons. Blocks, &c. Description.	Ship of 1100 tons. Blocks, &c. Size in Ins.	Ship of 1100 tons. Blocks, &c. Number.	Ship of 800 tons. Size of Rope in Inches.	Ship of 800 tons. Blocks, &c. Description.	Ship of 800 tons. Blocks, &c. Size in Ins.	Ship of 800 tons. Blocks, &c. Number.	Ship of 600 tons. Size of Rope in Inches.	Ship of 600 tons. Blocks, &c. Description.	Ship of 600 tons. Blocks, &c. Size in Ins.	Ship of 600 tons. Blocks, &c. Number.	Barque of 400 tons.	Barque of 300 tons.	Brig of 200 tons.
MIZEN-TOPGALLANT-YARD.															
Jackstays (iron)	1¾				1½				1½						
Foot-ropes	1¾				1¾				1¾						
Parral-lashing	2				2				2						
Lifts	2¼	T.B.	5	2	2¼	T.B.	6	2	2	T.B.	5	2			
Halliards and strapping	1¾				1¾				1¾						
Sheets	1¾	T.		2	1¾	T.	5	2	1½	T.	5	2			
Clue-lines	1¾				1¾				1¾						
Bow-lines and strapping	1¾	D.		2	1¾	D.	4	2	1¾	D.	4	2			
Bridles															
Earings	3				2				1¾						
Strapping, quarter-blocks		S.	6	2		S.	5	2		S.	4	2			
MIZEN-ROYAL-YARD.															
Jackstays (iron)	1				1				1						
Foot-ropes	1				1				1						
Braces and strapping	1¾	S.	4	2	1¾	S.	4	2	1¾	S.	4	2			
Parral-lashing	2				1				1						
Lifts	1				1				1						
Halliards	●														
Sheets		S.		2		S.		2		S.		2			
Clue-lines and strapping															
Earings, mantise															

SPANKER BOOM.																								
Topping-lifts	4½	S	13	2	4	S	13	2	4	S	11	2	3½	S	10	2	3	S	9	2	4	S	12	2
Falls and strapping . .	3	D	9	2	3	D	9	2	2½	D	8	2	2	D	7	2	2	D	7	2	2	D	7	2
		S*	9	2		S*	9	2		S*	8	2		S*	7	2		S*	7	2		S*	7	2
Boom-sheet	3½	D	11	2	3½	D	11	2	3	D	10	1	3	D	10	1	2½	D	8	1	3	D	10	4
										S	10	1		S	10	1		S	8	1				
Outhauller	3½	C	7	1	3	C	6	1	3	C	6	1	2½	C	5	1	2	C	5	1	2½	C	7	1
Guy-pendants . . .	3½	..	..	..	3½	..	..	..	3	..	..	..	3	..	..	..	3	..	..	..	4	*	..	1
																						T	..	1
Falls and strapping . .	2½	D	8	2	2	D	7	2	2	D	7	2	2	D	7	2	1½	D	5	2	2	D*	7	2
		S*	8	2		S*	7	2		S*	7	2		S*	7	2		S*	5	2		S*	7	2
GAFF.																								
Throat-Halliards . . .	4	D	12	1	3½	D	11	1	3	D	9	1	2½	D	9	1	2½	S	9	1	2½	D.I.bd	9	1
		T	..	1		T	..	1		T	..	1		T	..	1		T	..	1				
Peak-halliards and strappg.	4	D.I.bd	12	1	3½	D.I.bd	11	1	3	D.I.bd	10	1	3	D.I.bd	10	1	2½	D.I.bd	16	1	2½	S.I.bd	9	4
		S	12	2		S	11	2		S	10	2		S	10	2		S	16	2				
Vang-pendants . . .	3½	..	..	..	3	..	..	..	2½	..	..	..	2½	..	..	..	2½	..	..	..	..	..	..	..
Falls and strapping . .	2	D	7	2	2	D	7	2	2	D	7	2	2	D	7	2	2	D	7	2	..	..	..	..
		S*	7	2		S*	7	2		S*	7	2		S*	7	2		S*	7	2	..	..	..	..
Peak-brails	2	S	7	2	2	S	7	2	2	S	7	2	1½	S	6	2	1½	S	6	2	..	..	..	..
Throat-brails	2½	Tr.	8	2	2½	Tr.	8	2	2	Tr.	7	2	2	Tr.	7	2	2	Tr.	7	2	2½	C	6	2
Middle-brails	2	D	7	2	2	D	7	2	2	D	7	2	1½	D	6	2	1½	D	6	2	..	..	..	..
Hook-brails	2	S	7	2	1½	S	6	2	1½	S	6	2	1½	S	6	2	1½	S	6	2	..	..	..	..

EXPLANATIONS OF THE ABBREVIATIONS IN THE PRECEDING AND FOLLOWING TABLES.

C	Clump-block.	I.B.S. . .	Iron-bound Single-block.	St. bd. . .	Strap-bound-block.
D	Double-block.	I.B.S.C. . .	Iron-bound Single Clump-block.	T. . . .	Thimble.
D.E. . .	Dead-eye.	I.B.D. . .	Iron-bound Double-block.	Tr. . . .	Treble-block.
H. . . .	Heart.	Pl. D.E. . .	Plates with Dead Eyes.	I.B.D.C. . .	Iron-bound Double Clump-block.
H. & T., or * .	Hook and Thimble.	S. . . .	Single-block.	Double . .	Double-block.
I. bd. . .	Iron-bound.	Sis. . . .	Sister-block.	Single. . .	Single-block.

TABLE SHOWING THE SIZE OF THE RIGGING FOR SCHOONERS.

Names of Rigging.	180 to 200 tons. Size in inches.	Description of Blocks, &c.	Number	Inches.	Hooks.	Thimbles.	170 tons. Size in inches.	Description of Blocks, &c.	Number.	Inches.	Hooks.	Thimbles.	100 to 130 tons. Size in inches.	Description of Blocks, &c.	Number.	Inches.	Hooks.	Thimbles.
Bowsprit.																		
Gammoning, iron clamp																		
Shrouds, chain	4						3						4$\frac{5}{16}$					
Bobstays, chain	3						3						3$\frac{1}{16}$					
Jib-boom.																		
Jib-stay	5	D. I. B.	2	7	2	2	4$\frac{3}{4}$	D. I. B.	2	7	2	2	4$\frac{1}{4}$	D. I. B.	2	7	2	2
Purchase	2$\frac{1}{4}$	Single C.	2	7			2$\frac{1}{2}$	Single C.	2	7			2	Single C.	2	9		
Guys	2		2				2$\frac{1}{4}$						4$\frac{3}{4}$		2	10		
Runners	3$\frac{1}{2}$	B. I. B.	2	7	2		3$\frac{1}{4}$	B. I. B.	2	7			3	B. I. B.	2			
Falls	2$\frac{1}{2}$	Fiddle	2	12			2$\frac{1}{2}$	Fiddle	2	11			2	Fiddle	2	6		
Martingale-stay, chain	4						1$\frac{7}{16}$						3$\frac{3}{4}$					
Back-ropes	4$\frac{1}{4}$	Double	2	7	2		4$\frac{1}{4}$	Double	2	7			3$\frac{3}{4}$	Double	2	7		
Falls	2$\frac{3}{4}$	B. I. B.	2	7		1	2$\frac{1}{4}$	B. I. B.	2	8		1	2$\frac{3}{4}$	B. I. B.	2	7	1	1
Foot-ropes	2$\frac{1}{4}$						2$\frac{1}{4}$		2				2$\frac{1}{4}$		2	7	2	2
Heel-rope	3$\frac{1}{4}$	S. I. B.			1		3	S. I. B.	2		1	1	3	S. I. B.	3		1	
Jib-halliards	2$\frac{1}{4}$	Single	1	9			3$\frac{1}{4}$	Single	1	8			2$\frac{3}{4}$	Single	2	6	1	
Tack, traveller	2$\frac{1}{4}$		2	8			2$\frac{1}{4}$		2	7	1	2	2$\frac{1}{4}$		2	7		
Downhaul	3$\frac{1}{4}$	Single					2$\frac{1}{4}$	Single					3	Single				
Outhaul	4$\frac{1}{4}$	Single					4	Single					3$\frac{1}{4}$	Single				
Sheet-pendants	2$\frac{1}{4}$	Single					2$\frac{1}{4}$	Single					2$\frac{1}{4}$	Single				
Sheets					1	1					1	1					1	1
Jib-top-sail halliards	1$\frac{1}{2}$				1	2	1$\frac{1}{2}$		1		1	1	1$\frac{1}{2}$		1		1	1
Tack	2$\frac{1}{4}$						2$\frac{1}{4}$		2		2	2	3$\frac{3}{4}$		2		2	2
Sheets																		

FOREMAST.																		
Shrouds and pendants	6½	Dead-Eyes	6	8	..	..	6	Dead-Eyes	6	7½	..	..	5½	Dead-Eyes	6	7	..	..
		Single C.	2	8	..	..		Single C.	2	8	..	..		Single C.	2	7	..	..
Ratlines	9-thr	.. ..	..	..	..	..	9-thr	.. ..	..	..	..	..	9-thr	.. ..	..	..	..	..
Runners of tackles	4½	.. ..	..	..	2	2	4½	.. ..	..	..	2	2	4½	.. ..	..	..	2	2
Falls	2½	Double	2	8	..	..	2½	Double	2	7	..	..	2½	Double	2	7	..	..
		S. I. B.	2	8	..	..		S. I. B.	2	7	..	..		S. I. B.	2	7	..	..
Fore-stay	9	.. ..	..	..	..	..	3½	.. ..	..	..	..	..	7	.. ..	..	..	..	..
Laniard	2½	.. ..	..	..	..	..	2	.. ..	..	..	..	..	2	.. ..	..	..	..	..
Storm-stay	4½	.. ..	..	..	..	..	4½	.. ..	..	..	..	..	3½	.. ..	..	..	..	..
Laniard	2	.. ..	..	..	..	..	2	.. ..	..	..	..	..	1¾	.. ..	..	..	..	..
Lacing	2½	.. ..	..	..	..	..	2	.. ..	..	..	..	..	1¾	.. ..	..	..	..	..
Halliard	2¾	S. I. B.	2	9	..	..	2½	S. I. B.	2	8	..	..	2½	S. I. B.	2	7	..	..
Tack	3¾	.. ..	..	..	1	1	3½	.. ..	..	..	1	1	2¾	.. ..	..	..	1	1
Fall	2½	D. I. B.	1	7	1	1	2	D. I. B.	1	7	1	1	1¾	D. I. B.	1	6	1	1
		S. I. B.	1	7	..	..		S. I. B.	1	7	..	..		S. I. B.	1	6	..	..
Downhaul	2½	Single	1	7	..	..	2	Single	1	7	..	..	1¾	Single	1	6	..	..
Sheets	2¾	Double	2	9	..	..	2½	Double	2	8	..	..	2½	Double	2	7	..	..
		D. I. B.	2	9	..	..		D. I. B.	2	8	..	..		D. I. B.	2	7	..	..
FORE-YARD.																		
Square-sail halliards	2½	D. I. B.	2	8	..	..	2	D. I. B.	2	7	..	..	2	D. I. B.	2	6	..	..
Braces	2½	S. I. B.	2	6	4	..	2	S. I. B.	2	5	4	..	1¾	S. I. B.	2	5	4	..
Lifts	2½	.. ..	..	..	..	..	2	.. ..	..	..	..	..	2	.. ..	..	..	..	..
Yard-ropes	3½	Double	2	9	..	..	3	Double	2	8	..	..	2¾	Double	2	7	..	..
TOP-SAIL-YARD.																		
Square top-sail sheets	2½	Double	1	7	..	..	2½	Double	1	6	..	..	2	Double	1	6	..	..
Halliards	3	.. ..	..	..	..	..	2¾	.. ..	..	..	..	..	2½	.. ..	..	..	..	..
FORE-TOP-MAST.																		
Shrouds	3½	.. ..	..	..	..	..	3	.. ..	..	..	..	..	2¾	.. ..	..	..	..	..
Stay	3½	.. ..	..	..	..	..	3	.. ..	..	..	..	..	2¾	.. ..	..	..	..	..
Tackle	1¾	Double	1	6	..	..	1¾	Double	1	6	..	..	1½	Double	1	5	..	..
		S. I. B.	1	6	..	..		S. I. B.	1	6	..	..		S. I. B.	1	5	..	..

* Square-sail Halliards always used.

TABLE SHOWING THE SIZE OF THE RIGGING FOR SCHOONERS.

Names of Rigging.	180 to 200 tons.						170 tons.						100 to 150 tons.					
	Size in Inches.	Description of Blocks, &c.	Number.	Inches.	Hooks.	Thimbles.	Size in Inches.	Description of Blocks, &c.	Number.	Inches.	Hooks.	Thimbles.	Size in Inches.	Description of Blocks, &c.	Number.	Inches.	Hooks.	Thimbles.
Fore-Top-Mast—*continued.*																		
Backstays	3½	.. ..	..	..	..	2	3	.. ..	..	..	..	2	2¾	.. ..	..	..	..	2
Tackle	1¾	D. I. B.	4	6	..	..	1¾	D. I. B.	4	6	..	..	1½	D. I. B.	4	5	..	..
Mast-rope	2¾	.. ..	..	..	..	..	2½	.. ..	..	..	..	..	2½	.. ..	..	..	..	..
Gaff Fore-sail.																		
Throat halliards	3¾	D. I. B.	2	10	..	..	3½	D. I. B.	2	9	..	..	3	D. I. B.	2	8	..	..
Tricing line	2	Single	2	7	2	..	2	Single	2	7	2	..	1¾	Single	2	6	2	..
Peak halliards	3¾	S. I. B.	5	10	..	..	3½	S. I. B.	5	9	..	..	3	S. I. B.	3	8	..	..
Purchase	2	S. I. B.	1	7	..	..	1¾	S. I. B.	1	6	..	..	..	.. ..	..	..	..	..
		Double	1	7	..	..		Double	1	6	..	..						
Downhaul	1¾	Single	1	6	..	..	1¾	Single	1	6	..	..	1¾	Single	1	6	..	..
Fore-sheets	3	D. I. B.	2	9	..	..	2¾	D. I. B.	2	8	..	..	2½	D. I. B.	2	7	..	..
		Double	2	9	..	..		Double	2	8	..	..		Double	2	7	..	..
Fore-Gaff-Top-sail.																		
Halliards	3	Traveller	..	..	..	..	2¾	Traveller	..	..	..	..	2½	Traveller	..	..	..	..
Sheet	2½	Single	1	7	..	..	2¼	Single	1	7	..	..	2	Single	1	6	..	..
Tackle	1¾	Double	1	6	..	..	1¾	Double	1	6	..	..	1½	Double	1	5	..	..
		S. I. B.	1	6	..	..		S. I. B.	1	6	..	..		S. I. B.	1	5	..	..
Downhaul	1½	Single	1	5	..	..	1½	Single	1	5	..	..	1½	Single	1	5	..	..
Main-Mast.																		
Shrouds	6½	Dead-Eyes	4	8	..	..	6	Dead-Eyes	4	7	..	..	5½	Dead-Eyes	4	7	..	..
Pendants	5	Single C.	2	8	..	..	4¾	Single C.	2	8	..	..	4¼	Single C.	2	7	..	..

Runners	4	2½	6½	4½	2¾					
Falls						D.I.B.				
Jumper stays										
Runners										
Tackles										
BOOM-MAIN-SAIL.										
Main halliards	3½	3½	3	1½	2	2	4½	1	D.I.B. / S.I.B. / Double / S.I.B. / Single / D.I.B. / Single		
Peak halliards										
Purchase										
Downhaul										
Tack Tackle										
Tack tricing-line										
Reef-earings										
Lacing										
MAIN-BOOM.										
Topping-lifts	5	2½	3½	2½	4	2½	B.I.B.C. / Double / S.I.B. / Double / Single C. / Fiddle / S.I.B. / Double / Single		
Tackle-falls										
Boom-sheets										
Reef-tackle										
Boom-guy-pendant										
Guy-tackle										
MAIN-TOP-MAST.										
Shrouds	3½	3½	1¾				D.I.B.			
Backstays										
Tackles										

TABLE SHOWING THE SIZE OF THE RIGGING FOR SCHOONERS.

Names of Rigging.	180 to 200 tons.						170 tons.						100 to 130 tons.					
	Size in Inches.	Description of Blocks, &c.	Number.	Inches.	Hooks.	Thimbles.	Size in Inches.	Description of Blocks, &c.	Number.	Inches.	Hooks.	Thimbles.	Size in Inches.	Description of Blocks, &c.	Number.	Inches.	Hooks.	Thimbles.
Main-Top-Mast.—*continued.*																		
Jumper-stays	3¼	.. ..	..	..	..	..	3	.. ..	..	..	..	..	2¾	.. ..	..	..	..	..
Tackles	1¾	Single	4	6	..	..	1¾	Single	4	6	..	..	1½	Single	4	5	..	..
Mast-rope	3	.. ..	..	..	..	..	2¾	.. ..	..	..	..	..	2½	.. ..	..	..	..	..
Main-Gaff-Top-sail.																		
Halliards	3	Traveller	..	..	..	..	2¾	Traveller	..	..	..	..	2½	Traveller	..	..	..	..
Sheets	2½	Single	1	7	..	..	2½	Single	1	7	..	..	2	Single	1	6	..	..
Tack	2½	.. ..	..	..	1	1	2½	.. ..	..	..	1	1	2	.. ..	..	..	1	1
Tackle	1¾	{ Double	1	6	..	..	1¾	{ Double	1	6	..	..	1½	{ Double	1	5	..	..
		{ S. I. B.	1	6	..	..		{ S. I. B.	1	6	..	..		{ S. I. B.	1	5	..	..
Downhaul	1½	Single	1	5	..	..	1½	Single	1	5	..	..	1½	Single	1	5	..	..

TABLE SHOWING THE SIZE OF THE RIGGING FOR CUTTER YACHTS

Names of Rigging.	30 to 40 tons.						50 to 60 tons.						70 to 90 tons.					
	Size in Inches.	Description of Blocks, &c.	Number.	Inches.	Hooks.	Thimbles.	Size in Inches.	Description of Blocks, &c.	Number.	Inches.	Hooks.	Thimbles.	Size in Inches.	Description of Blocks, &c.	Number.	Inches.	Hooks.	Thimbles.
Bowsprit.																		
Shrouds, *wire*	1¼	I. B. D.	2	6	2	2	1¾	D. I. B.	4	7	2	2	1⅞	D. I. B.	4	8	2	2
		I. B. S.	2	6	..	..												
Falls	1½	.. ..	..	..	..	..	2¼	.. ..	..	..	..	..	2½	.. ..	..	..	..	..
Bobstay pendants, *wire*	1¾	I. B. D.	1	6	1	1	2	Shackle	..	..	..	..	2⅜	Shackle	..	..	..	..
		I. B. S.	1	6	..	..												
Fall	2	.. ..	..	..	..	..	8	D. I. B.	1	8	1	1	3½	D. I. B.	1	9	1	1
								S. I. B.	1	8	..	..		S. I. B.	1	9	..	..
Heel-rope	2¼	S. I. B.	1	6	1	1	8	S. I. B.	1	8	1	1	3½	S. I. B.	1	9	1	1
Jib-tack	8	.. ..	..	..	..	..	4	.. ..	..	..	..	..	4½	.. ..	..	..	..	..
Whip	1½	Single	1	6	..	..	2½	Single	1	7	..	..	2¾	Single	1	8	..	..
Halliards	3	S. I. B. C.	1	6	..	..	4	S. I. B. C.	1	9	..	..	4½	S. I. B. C.	1	10	..	..
		Single C.	1	6	..	..		Single C.	2	9	..	..		Single C.	2	10	..	..
Purchase	1½	S. I. B.	1	5	..	..	1¾	D. I. B.	2	7	..	..	2	D. I. B.	2	8	..	..
		Single	1	5	..	..												
Downhaul	1	.. ..	..	..	..	..	1½	Single	1	5	..	..	1½	Single	1	5	..	..
Inhaul	1¼	.. ..	..	..	..	..	1½	.. ..	..	..	..	..	1¾	.. ..	..	..	..	..
Sheets	3½	.. ..	..	..	..	..	2½	Single C.	2	7	..	..	2½	Single C.	2	8	..	..
Bobstay tricing-line	1	.. ..	..	..	..	..	1¼	Single	1	5	..	..	1½	Single	1	5	..	..
Mast.																		
Shrouds, *wire*	1⅞	Dead-Eyes	6	5	..	..	2⅛	Dead-Eyes	8	6½	..	..	2⅜	Dead-Eyes	8	8	..	..
Laniards	2¾	.. ..	..	..	..	..	3	.. ..	..	..	..	..	3¼	.. ..	..	..	..	..
Mast-head pendants, *wire*	1⅞	Single C.	2	6	..	..	1⅝	S. I. B. C.	2	8	..	..	1⅞	S. I. B. C.	2	10	..	..
Runners	3	Single	2	5	2	2	4	Fiddle	2	11	2	2	4½	Fiddle	2	12	2	2
Falls	1¾	S. I. B.	2	5	2	2	2	Single	2	7	2	2	2½	Single	2	8	2	2
Stay	2⅛	.. ..	..	..	..	1	3½	Heart	1	8	..	..	4	Heart	1	9	..	..
Laniard	2	.. ..	..	..	..	..	2¼	.. ..	..	..	..	..	2½	.. ..	..	..	..	..

TABLE SHOWING THE SIZE OF THE RIGGING FOR CUTTER YACHTS.

Names of Rigging.	30 to 40 tons. Size in Inches.	Description of Blocks, &c.	Number.	Inches.	Hooks.	Thimbles.	50 to 60 tons. Size in Inches.	Description of Blocks, &c.	Number.	Inches.	Hooks.	Thimbles.	70 to 90 tons. Size in Inches.	Description of Blocks, &c.	Number.	Inches.	Hooks.	Thimbles.
Mast—*continued.*																		
Fore-halliard	1½	S. I. D.	2	5	..	..	2½	S. I. B.	2	6	..	..	2¾	S. I. B.	2	8	..	..
Downhaul	1½	.. ..	..	..	..	..	1¾	.. ..	..	..	..	..	2	.. ..	..	..	..	..
Tack	2	.. ..	..	..	..	..	2½	.. ..	..	..	..	..	3	.. ..	..	..	..	..
Fall	..	.. ..	..	..	..	..	1½	S. I. B.	2	5	..	..	1¾	D. I. B.	2	5	..	..
Sheets	1¾	S. I. B.	2	5	2	4	2	D. I. B.	2	6	2	4	2¼	D. I. B.	2	7	2	4
		Single	2	5	..	..												
Jack-stay (*iron*)	..	.. ..	..	..	..	..	..	.. ..	..	..	..	..	..	.. ..	..	..	..	..
Square-Sail-Yard.																		
Square-sail braces	1	.. ..	..	..	..	..	1½	S. I. B.	2	5	..	..	1¾	S. I. B.	2	6	..	..
Halliards	1¼	S. I. B.	2	4	..	..	2	S. I. B.	2	5	..	..	2¼	S. I. B.	2	6	..	..
Sheets and guys	1½	.. ..	..	..	..	..	2	.. ..	..	..	..	..	2½	.. ..	..	..	..	..
Yard-ropes	1½	Single	2	5	2	2	2¼	Single	2	6	2	2	3	Single	2	8	2	2
Boom-Main-Sail.																		
Throat halliards	2½	D. I. B.	1	5	..	..	3¼	D. I. B.	2	9	..	..	3½	D. I. B.	2	10	..	..
		S. I. B.	1	5	..	..												
Peak halliards	2½	S. I. B.	5	6	..	..	3¼	S. I. B.	5	9	..	..	3½	S. I. B.	5	10	..	..
Purchase	..	.. ..	..	..	..	..	1¾	D. I. B.	1	7	..	..	1¾	D. I. B.	1	7	..	..
								S. I. B.	1	7	..	..		S. I. B.	1	7	..	..
Tack-tackle	1	Single	2	4	2		1½	Double	1	6	1	1	1¾	Double	1	6	..	..
						2		S. I. B.	1	6	..	..		Single	1	6	..	..
Tricing-line	1	Single	2	5	..	..	1¾	Single	2	6	..	..	1¾	Single	2	6	..	..
Peak downhaul	1	Single	1	4	..	..	1½	Single	1	5	..	..	1¾	Single	1	5	..	..
Reef pendants	2	.. ..	..	..	..	..	3½	.. ..	..	..	..	..	4	.. ..	..	..	..	..
Lacing and earings	1	.. ..	..	..	..	..	1½	.. ..	..	..	..	..	1¾	.. ..	..	..	..	..
Tackle	1	S. I. B.	1	5	..	..	1¾	Fiddle	1	7	..	..	2	Fiddle	1	10	..	..
		Single	1	5	..	..		Single	1	6	..	..		Single	1	6	..	..

Boom-guy pendants	2¾	.. ..	..	..	1	2	4	.. ..	..	..	1	2	4½	.. ..	..	..	1	2
Tackle	1½	Single	2	5	2	2	1¾	{ Double	1	7	..	..	2	{ Double	1	7	..	..
								{ Single	1	7	..	..		{ Single	1	7	..	..
Topping-lifts, *wire*	1⅛	Single	2	5	2	2	1⅝	S. I. B.	4	7	..	..	1⅝	S. I. B.	4	7	..	..
Fall	1½	S. I. B.	2	5	..	..	2	S. I. B.	4	7	..	..	2½	{ D. I. B.	4	8	..	..
														{ S. I. B.	4	8	..	..
Main-sheet	2	{ Double	1	7	..	..	3	{ Double	1	10	..	..	3½	{ Double	1	11	..	..
		{ S. I. B. C.	1	7	..	..		{ D. I. B.	1	10	..	..		{ D. I. B.	1	11	..	..
								{ S. I. B. C.	1	7	..	..		{ S. I. B. C.	1	7	..	..
Try-sail-sheets	1½	{ D. I. B.	2	7	..	..	2¼	D. I. B.	4	8	4	..	2½	D. I. B.	4	8	4	..
		{ S. I. B.	2	7	..	..												
Downhaul	1	Single	1	4	..	..	1½	Single	1	5	..	..	1¾	Single	1	5	..	..
TOP-MAST.																		
Shrouds, *wire*	1	.. ..	..	..	..	2	1½	.. ..	..	..	..	4	1½	.. ..	..	..	..	4
Tackle	1	S. I. B.	4	4	..	..	1½	D. I. B.	8	5	..	..	1¾	D. I. B.	8	6	..	..
Stay, *wire*	1	S. I. B. C.	1	5	..	..	1½	S. I. B. C.	1	5	..	..	1⅝	S. I. B. C.	1	5	..	..
Tackle	..	.. ..	..	..	..	..	1½	{ Double	1	5	..	..	1¾	{ Double	1	5	..	..
								{ S. I. B.	1	5	..	..		{ S. I. B.	1	5	..	..
Mast-rope	2	.. ..	..	..	..	..	3	.. ..	..	..	..	..	3¾	.. ..	..	..	..	..
Tackle	..	.. ..	..	..	..	..	1½	{ D. I. B.	1	5	..	..	1¾	{ D. I. B.	1	6	..	..
								{ S. I. B.	1	5	..	..		{ S. I. B.	1	6	..	..
Gaff-top-sail halliards	2	.. ..	..	..	..	..	..	.. ..	..	..	..	..	..	.. ..	..	..	..	..
Tye	..	.. ..	..	..	..	..	3	Single	1	6	..	..	3¼	Single	1	6	..	..
Whip	..	.. ..	..	..	..	..	1½	.. ..	..	..	1	..	2¼	.. ..	..	..	1	..
Clue-line	1½	Single	1	4	..	..	1½	Single	1	5	..	..	1¾	Single	1	5	..	..
Sheet	1¾	.. ..	..	..	1	..	2½	Single C.	1	5	..	..	3	Single C.	1	5	..	..
Tack	1⅛	.. ..	..	..	..	..	2½	.. ..	..	..	..	..	2¾	.. ..	..	..	..	..
Tackle	..	.. ..	..	..	..	..	1½	{ D. I. B.	1	6	..	..	1¾	{ D. I. B.	1	6	..	..
								{ S. I. B.	1	6	..	..		{ S. I. B.	1	6	..	..
Jib-top-sail halliards	1¾	.. ..	..	..	..	..	2¼	.. ..	..	..	..	..	2½	.. ..	..	..	..	..
Downhaul	..	.. ..	..	..	..	..	1¾	Single	1	5	..	..	1¾	Single	1	5	..	..
Tack	1	Single	1	5	..	..	1¾	Single	2	5	..	..	1¾	Single	2	5	..	..
Sheets	1	.. ..	..	..	..	..	1¾	.. ..	..	..	..	1	1¾	.. ..	..	..	..	..
Signal halliards	1½	.. ..	..	..	..	..	½	.. ..	..	..	..	..	¾	.. ..	..	..	..	..
Half-top-sail halliards	1	.. ..	..	..	..	..	2¼	.. ..	..	..	..	..	2½	.. ..	..	..	..	..
Downhaul	1½	.. ..	..	..	..	..	1½	Single	1	5	..	..	1¾	Single	1	5	..	..
Tack	1	.. ..	..	..	2	..	1½	.. ..	..	..	2	..	1¾	.. ..	..	..	2	..

A TABLE SHOWING THE COMPARATIVE SIZES, WEIGHT, AND STRENGTH BETWEEN NEWALL AND CO.'S PATENT WIRE ROPES AND HEMP ROPES FOR STANDING RIGGING.

Hemp Rope.		Wire Rope of Equivalent Strength.			
Circumference.	lb. Weight per fathom.	Circumference.	lb. Weight per fathom.	Breaking Strain.	Working Load.
2¾	2	1	1	2 tons.	6 cwt.
		1¼	1½	3 ,,	9 ,,
3¾	4	1⅜	2	4 ,,	12 ,,
		1¾	2½	5 ,,	15 ,,
4½	5	1⅞	3	6 ,,	18 ,,
		2	3½	7 ,,	21 ,,
5½	7	2⅛	4	8 ,,	24 ,,
		2¼	4½	9 ,,	27 ,,
6	9	2⅜	5	10 ,,	30 ,,
		2½	5½	11 ,,	33 ,,
6½	10	2⅝	6	12 ,,	36 ,,
		2¾	6½	13 ,,	39 ,,
7	12	2⅞	7	14 ,,	42 ,,
		3	7½	15 ,,	45 ,,
7½	14	3⅛	8	16 ,,	48 ,,
		3¼	8½	17 ,,	51 ,,
8	16	3⅜	9	18 ,,	54 ,,
		3⅜	9½	19 ,,	57 ,,
8½	18	3½	10	20 ,,	60 ,,
		3⅝	11	22 ,,	66 ,,
9½	22	3¾	12	24 ,,	72 ,,
		3⅞	13	26 ,,	78 ,,
10	26	4	14	28 ,,	84 ,,

A TABLE SHOWING THE COMPARATIVE STRENGTH BETWEEN IRON CHAINS AND HEMP ROPES.

Size of the Chains.	Weight in lb. per fathom.	Proof Strain in tons.	Size of Rope.	Weight of Rope in lb. per fathom.
5/16	6	⅝	2½	1½
⅜	8½	1½	3½	2½
7/16	11	2½	4	3¾
½	14	3½	4¾	5
9/16	18	4½	5½	7
⅝	24	5¼	6¼	8½
11/16	28	6¼	7	10½
¾	32	7¾	7½	12
13/16	36	9¼	8¼	15
⅞	44	10¾	9	17½
15/16	50	12½	9½	19½
1	56	14	10	22

NOTE.—One-eighth of an inch of iron in diameter is more than equal to an inch of hemp rope in circumference.

A TABLE SHOWING THE STRENGTH OF SHORT ROUND-LINKED BOBSTAY, BOWSPRIT-SHROUD OR CRANE-CHAIN, WITHOUT STUDS, SUCH AS IS USED FOR RIGGING, ETC.

Size.	Weight of 100 fathoms in lb.	Breaking Strain in tons. Maximum.	Breaking Strain in tons. Minimum.	Mean.	Required Test of Strength.
1 5/8	15,500	75·	68·	73·	31·6
1 1/2	.. ..	64·	56·2	62·8	27·
1 7/16	.. ..	59·	53·8	57·4	24·7
1 3/8	.. ..	54·2	49·6	52·8	22·6
1 5/16	.. ..	47·7	45·5	48·4	20·6
1 1/4	.. ..	45·3	41·7	44·1	18·8
1 3/16	.. ..	41·2	38·	40·1	17·
1 1/8	7481	37·3	34·5	36·3	15·3
1 1/16	.. ..	33·6	31·2	32·7	13·6
1	6490	30·1	28·1	29·8	12·
15/16	5600	26·8	25·2	26·1	10·5
7/8	4500	23·7	22·5	23·1	9·1
13/16	4000	20·9	20·	20·4	7·9
3/4	3449	17·8	16·6	17·3	6·8
11/16	2900	14·9	13·5	14·6	5·6
5/8	2538	12·3	10·8	12·	4·6
9/16	2001	10·	8·7	9·7	3·8
1/2	1583	7·9	6·9	7·7	3·
7/16	1060	6·	5·2	5·9	2·3
3/8	827	4·4	3·8	4·3	1·6
5/16	581	3·	2·7	3·	1·1
1/4	392	1·9	1·7	1·9	·75
3/16	.. ..	1·1	·97	1·	·42

A TABLE SHOWING THE WEIGHT OF CHAIN CABLE.

150 Fathoms of 2¼ inch . . . Weighs . 45,249 pounds.
150 „ 2⅛ „ . . . . „ . 37,400 „
150 „ 2 „ . . . „ . 37,872 „
150 „ 1 15/16 „ . . . . „ . 34,125 „
150 „ 1⅞ „ . . . „ . 32,225 „
150 „ 1¾ „ . . . . „ . 27,102 „
150 „ 1 11/16 „ . . . „ . 25,350 „
150 „ 1⅝ „ . . . . „ . 23,934 „
150 „ 1⅜ „ . . . „ . 17,204 „
150 „ 1¼ „ . . . . „ . 14,884 „
150 „ 1⅛ „ . . . „ . 11,921 „

CORDAGE TABLE,* SHOWING THE WEIGHT OF ONE FATHOM OF ROPE, FROM 1 INCH TO 24 INCHES INCLUSIVE, PLAIN LAID THREE-STRAND, SUCH AS USED FOR RUNNING RIGGING, ETC.

Size of Rope in inches.	Weight per fathom. lb.	oz.	Size of Rope in inches.	Weight per fathom. lb.	oz.	Size of Rope in inches.	Weight per fathom. lb.	oz.
1	0	3¾	6¼	9	0	11½	30	9
1¼	0	5¾	6½	9	11¾	11¾	31	14
1½	0	8¼	6¾	10	8	12	33	3
1¾	0	11¼	7	11	4⅝	12¼	34	9½
2	0	14¾	7¼	12	2	12½	36	0
2¼	1	2⅝	7½	13	0⅜	12¾	37	8
2½	1	7	7¾	13	13⅜	13	38	15
2¾	1	11⅞	8	14	12⅓	13¼	40	8¼
3	2	1⅝	8¼	15	11⅓	13½	42	0
3¼	2	7	8½	16	10½	13¾	43	9½
3½	2	13⅛	8¾	17	10⅓	14	45	4¾
3¾	3	2⅜	9	18	10⅝	15	52	0½
4	3	11	9¼	19	11⅜	16	59	5
4¼	4	1⅜	9½	20	13⅜	17	66	10
4½	4	10⅜	9¾	21	14⅜	18	74	10
4¾	5	3⅜	10	23	1½	19	83	2
5	5	12¼	10¼	24	3½	20	92	11
5¼	6	5⅝	10½	25	7	21	102	1
5½	7	0	10¾	26	11⅜	22	112	0
5¾	7	9⅞	11	27	14½	23	122	3
6	8	4¼	11¼	29	1½	24	134	6

* *Rule to find the Weight of any sized Rope.*—A rope of 1 inch circumference requires 486 fathoms to make one hundred weight. The superficial part of all circles being in proportion to the square of their diameters, consequently the square of their circumference. Therefore a rope of 1 inch in circumference, whose square is one, has 486 fathoms to a cwt.; and, therefore, 486 being divided by the square of the circumference of any rope, the quotient will give the number of fathoms. For instance—

$$9 \times 9 = 81)\ 486\ (6 \text{ the number of fathoms in a cwt.}$$
$$486$$

Rule to find the Weight of 120 fathoms of any sized Cable.—Multiply the circumference by the circumference and divide the product by 4, and the quotient will be the number of cwt. in 120 fathoms.

A TABLE OF THE WEIGHT OF TARRED CORDAGE.

Hawsers of 130 fathoms.				Hawsers of 120 fathoms.			
Size.	Weight per 130 fms. each.			Size.	Weight per 120 fms. each.		
Inches.	cwt.	qr.	lb.	Inches.	cwt.	qr.	lb.
6½	13	1	11	9½	22	2	9
6	11	1	13	9	20	1	17
5½	9	2	2	8½	18	0	26
5	7	3	19	8	16	0	6
4½	6	1	22	7½	13	3	16
4	5	0	14	7	12	0	18
3½	3	3	7	6½	10	1	19
3	2	3	20	6	9	0	12
2½	2	0	5	5½	7	3	7
2	1	1	6	5	6	2	1
1½	0	3	13	4½	5	0	23
1	0	1	20	4	4	0	18
¾	0	1	4	3½	3	1	22
..	..	..	..	3	2	2	11
..	..	..	..	2½	1	3	0
..	..	..	..	2	1	1	4

A TABLE SHOWING THE WEIGHT OF 100 FATHOMS OF CABLE-LAID ROPE, FROM 2 TO 24 INCHES.

ALSO A COMPARATIVE SIZE OF CHAIN.

Size.	Threads.	Weight.			Chain Equal.	Size.	Threads.	Weight.			Chain Equal.
		cwt.	qr.	lb.				cwt.	qr.	lb.	
2	27		3	26	..	13½	954	35	0	7	1¼
2½	36	1	1	8	..	14	1026	37	2	24	..
3	54	1	3	25	..	14½	1098	40	1	12	1⅜
3½	72	2	2	16	..	15	1170	43	0	1	..
4	99	3	1	6	⅝	15½	1251	45	3	26	1½
4½	108	3	3	24	..	16	1332	48	3	24	..
5	135	4	3	23	..	16½	1418	51	3	21	..
5½	162	5	3	22	..	17	1503	55	1	9	1⅝
6	189	6	3	21	..	17½	1593	58	2	6	..
6½	216	7	3	21	..	18	1683	61	3	13	1¾
7	252	9	1	1	..	18½	1782	65	2	1	..
7½	288	10	2	9	..	19	1881	69	0	17	1⅞
8	336	12	0	26	¾	19½	1980	72	3	4	..
8½	378	13	3	15	..	20	2033	76	3	1	..
9	423	15	2	25	..	20½	2187	80	1	16	..
9½	468	17	0	22	⅞	21	2295	84	1	14	2
10	522	19	0	21	1	21½	2403	88	1	10	..
10½	576	21	0	10	1	22	2520	92	2	16	..
11	630	23	0	18	..	22½	2646	97	1	3	..
11½	684	25	0	15	1⅛	23	2763	101	2	8	2⅛
12	747	27	1	23	1⅛	23½	2880	105	3	14	..
12½	810	29	3	3	..	24	3006	110	2	1	2¼
13	882	32	1	19	..						

A TABLE SHOWING THE STRENGTH OF PLAIN LAID ROPE OF THREE STRANDS.

·Size.	No. of Yarns in a Rope.	Weight of 100 fathoms in lb.	Breaking Strain in tons.		Mean.
			Maximum.	Minimum.	
12	1173	2940	45·5	35·	40·
11½	1077	..	41·7	32·	36·7
11	987	..	38·2	29·3	33·6
10½	900	..	34·9	26·7	30·7
10	816	2136	31·7	24·2	27·9
9½	738	..	28·6	21·8	25·2
9	660	1712	25·7	19·6	22·6
8½	591	..	23·	17·5	20·2
8	522	1379	20·4	15·5	18·
7½	450	..	18·	13·6	15·8
7	399	..	15·8	11·8	13·8
6½	345	..	13·7	10·2	12·
6	294	834	11·75	8·7	10·3
5½	249	712	9·8	7·3	8·7
5	204	..	8·2	6·1	7·2
4½	168	413	6·7	5·	5·9
4	132	..	5·3	4·	4·7
3½	102	..	4·1	3·2	3·7
3	75	203	3·1	2·4	2·8
2½	54	..	2·2	1·8	2·1
2	33	..	1·5	1·3	1·4
1¾	27	..	1·28	1·13	1·23
1½	21	..	·90	·86	·88
1¼	15	..	·60	·53	·56
1	12	..	·58	·46	·51
¾	9	..	·51	·42	·46
½	6	..	·28	·28	·28

APPENDIX.

DIMENSIONS OF MASTS AND YARDS OF HER MAJESTY'S SHIPS PHAETON AND VERNON OF 50 GUNS.

NAMES OF THE MASTS AND YARDS.	PHAETON. Ft. In.		VERNON. Ft. In.	
Main-mast				
Housing from the heel to the deck	29	4		
From deck to lower side of trestle-trees	64	6		
Head	19	10	116	0
Extreme length of mast	112	0	0	38
Diameter	0	37		
Main-top-mast				
Whole length	65	0	66	14
Diameter	0	21¼	0	21
Main-top-gallant-mast				
Diameter	0	13	0	11½
From lower side of fid-hole to hounds	29	0	29	0
Pole	19	0		
Main-yard				
Whole length, yard-arms included	96	0	96	8
Yard-arms, each	4	0		
Diameter	0	23	0	22¾
Main-topsail-yard				
Whole length, yard-arms included	68	0	74	0
Yard-arms, each	5	8		
Diameter	0	15	0	15
Main-top-gallant-yard				
Whole length, yard-arms included	43	0	45	10
Yard-arms, each	1	10		
Diameter	0	10	0	9
Royal-yard				
Whole length, yard-arms included	30	6		
Yard-arms, each	1	3		
Diameter	0	0		

N

	Measurement	ft in	ft in
Mizen-top-mast	Whole length, head included	46 6	48 10
	Head	6 3	0 .. 13½
	Diameter	0 14½	21 0
Mizen-topgallant-mast	From lower side of fid-hole to hounds	21 6	0 .. 8½
	Pole	14 0	
	Diameter	0 8¾	70 6
Mizen-crossjack-yard	Whole length, yard-arms included	64 2	0 .. 15
	Yard-arms, each	2 8	
	Diameter	0 16½	40 1
Mizen-topsail-yard	Whole length, yard-arms included	47 0	0 .. 9½
	Yard-arms, each	3 11	
	Diameter	0 10½	
Mizen-topgallant-yard	Whole length, yard-arms included	31 1	.. .. ..
	Yard-arms, each	1 4	.. .. ..
	Diameter	0 7½	
Mizen-royal-yard	Whole length, yard-arms included	23 0	.. .. ..
	Yard-arms, each	0 11	.. .. ..
	Diameter	0 4½	
Mizen-gaff	Length	45 0	40 7½
	Diameter	0 10½	0 10
Spanker-boom	Length	64 0	58 0
	Diameter	0 15	0 12¾
Bowsprit	Length, exclusive of housing	45 6	6 6
	Housing	9 7	0 34½
	Diameter	0 36	
Jib-boom	Whole length, housing included	47 0	48 0
	Housing	16 0	0 .. 15½
	Diameter	0 15½	
Flying-jib-boom	Length	60 8	.. ..
	Diameter	0 8¾	

These dimensions are also applicable for H.M. first-class frigates of 50 guns, Indefatigable, Leander, and Arethusa.

DIMENSIONS OF THE PRINCIPAL MASTS AND YARDS BELONGING TO THE SHIPS OF DIFFERENT RATES IN THE ROYAL NAVY OF GREAT BRITAIN AND IRELAND.

Right Divisional Dimensions of Masts and Yards	Brig of 10 guns				Brig of 18 guns				Frigate of 28 guns				Frigate of 46 guns			
	Masts		Yards		Masts		Yards		Masts		Yards		Masts		Yards	
	Length	Diameter	Length	Diameter	Length	Diameter	Length	Diameter	Length	Diameter	Length	Diameter	Length	Diameter	Length	Diameter
	Feet.	Feet.	Feet.	Feet.	Feet.	Feet.	Feet.	Feet.	Feet.	Feet.	Feet.	Feet.	Feet.	Feet.	Feet.	Feet.
Main-mast and yard	54·5	1·47	48·09	0·91	68·25	1·81	54·53	0·94	71·00	1·70	63·00	1·20	90·00	2·33	81·75	1·57
Top-mast and yard	31·0	0·83	37·5	0·66	38·09	1·02	41·00	0·70	43·16	1·00	40·50	0·81	54·00	1·34	59·00	1·13
Topgallant-mast and yard	19·5	0·50	26·0	0·5	25·08	0·59	27·50	0·50	21·53	0·59	28·33	0·50	27·00	0·75	37·50	0·60
Fore-mast and yard	46·5	1·33	43·00	0·91	58·75	1·00	54·50	0·94	64·50	1·58	55·00	1·05	82·50	2·08	71·41	1·35
Top-mast and yard	31·0	0·83	37·5	0·66	36·50	1·02	42·00	0·70	38·10	1·05	41·00	0·70	47·08	1·34	53·33	0·95
Topgallant-mast and yard	19·5	0·50	26·0	0·50	25·08	0·59	27·50	0·50	19·00	0·55	25·00	0·45	23·41	0·64	32·91	0·56
Mizen-mast	..	..	..	..	..	..	..	..	54·00	1·35	..	..	65·00	1·58	..	..
Cromjeck-yard	..	..	..	..	..	..	..	..	..	..	46·50	0·51	..	..	59·00	1·18
Top-mast and yard	..	..	..	..	..	..	..	..	32·41	0·75	31·00	0·48	41·00	0·99	40·66	0·63
Topgallant-mast and yard	..	..	..	..	..	..	..	..	16·33	0·45	21·33	0·37	20·50	0·57	28·00	0·45
Bowsprit	36·00	1·41	..	..	43·33	1·06	..	..	44·59	1·75	..	..	54·50	2·20	..	..
Spanker-boom	..	..	50·66	0·91	..	..	58·0	1·12	..	..	43·91	0·70	..	..	55·75	0·95
Gaff	..	..	30·33	0·82	..	..	34·0	0·73	..	..	32·50	0·64	..	..	30·00	0·73

DIMENSIONS OF THE PRINCIPAL MASTS AND YARDS BELONGING TO THE SHIPS OF DIFFERENT RATES IN THE ROYAL NAVY OF GREAT BRITAIN AND IRELAND.

Eight Divisional Dimensions of Masts and Yards	Frigate of 50 guns				Ship of 74 guns				Ship of 80 guns				Ship of 120 guns			
	Masts		Yards		Masts		Yards		Masts		Yards		Masts		Yards	
	Length	Diameter	Length	Diameter	Length	Diameter	Length	Diameter	Length	Diameter	Length	Diameter	Length	Diameter	Length	Diameter
	Feet.	Feet.	Feet.	Feet.	Feet.	Feet.	Feet.	Feet.	Feet.	Feet.	Feet.	Feet.	Feet.	Feet.	Feet.	Feet.
Main-mast and yard	114·0	3·00	96·66	1·80	108·0	3·00	96·66	1·89	118·08	3·31	103·25	2·06	119·66	3·33	104·33	2·05
Top-mast and yard	65·08	1·75	68·00	1·25	64·03	1·00	70·50	1·21	69·00	1·73	74·25	1·33	68·03	1·71	73·66	1·29
Topgallant-mast and yard	33·00	0·91	45·83	0·77	33·00	0·91	45·88	0·77	34·50	0·97	46·06	0·78	34·41	0·97	48·75	0·83
Fore-mast and yard	108·5	2·92	84·33	1·88	98·5	2·00	84·33	1·63	108·00	3·00	89·75	1·70	110·33	3·06	91·08	1·70
Top-mast and yard	58·66	1·75	62·00	1·10	57·66	1·00	61·50	1·00	62·16	1·73	64·50	1·15	63·08	1·71	64·50	1·14
Topgallant-mast and yard	29·08	0·80	40·00	0·70	29·08	0·80	40·00	0·70	30·00	0·83	33·43	0·66	31·00	0·66	42·66	0·71
Mizen-mast	92·05	2·00	··	··	73·09	1·80	··	··	81·66	2·06	··	··	81·00	2·04	··	··
Cross-jack-yard	··	··	70·50	1·21	··	··	70·50	1·21	··	··	74·25	1·33	··	··	74·50	1·29
Top-mast and yard	47·08	1·08	46·08	0·80	47·06	1·08	40·08	0·80	50·33	1·16	49·00	0·84	49·41	1·14	49·00	0·84
Topgallant-mast and yard	22·00	0·66	31·75	0·51	24·00	0·66	31·75	0·51	24·66	0·70	34·00	0·53	24·66	0·71	30·25	0·54
Bowsprit	66·00	2·92	··	··	66·0	2·92	··	··	71·09	3·00	··	··	63·00	3·07	··	··
Spanker-boom	··	··	64·25	1·28	··	··	68·25	1·06	··	··	70·01	1·15	··	··	60·58	1·14
Gaff	··	··	45·91	0·90	··	··	50·91	0·96	··	··	63·0	1·00	··	··	61·50	1·05

LONDON:
BRADBURY AND EVANS, PRINTERS, WHITEFRIARS.

Now in the course of Publication,

GREEK AND LATIN CLASSICS.

PRICE ONE SHILLING PER VOLUME,

Except in Some Instances, and those are at 1s. 6d. or 2s. each

VERY NEATLY PRINTED ON GOOD PAPER,

A SERIES OF VOLUMES

CONTAINING THE

PRINCIPAL GREEK AND LATIN AUTHORS,

ACCOMPANIED BY

EXPLANATORY NOTES IN ENGLISH, PRINCIPALLY SELECTED FROM THE BEST AND MOST RECENT GERMAN COMMENTATORS,

AND COMPRISING

All those Works that are essential for the Scholar and the Pupil, and applicable for use at the Universities of Oxford, Cambridge, Edinburgh, Glasgow, Aberdeen, and Dublin,—the Colleges at Belfast, Cork, Galway, Winchester, and Eton, and the great Schools at Harrow, Rugby, &c.,— also for Private Tuition and Instruction, and for the Library.

Vols. 1. 2, 3, 4, 5, 6, 7, 8, 9, 16, 17 of the Latin Series have appeared. Of the Greek Series, vols. 1, 2, 3, 4, 5, 6, 9, 18, 41 also have been published, and will be regularly continued.

LATIN SERIES.

1. A New LATIN DELECTUS, OR INTRODUCTORY CLASSICAL READER, consisting of Extracts from the best Authors, systematically arranged; accompanied by Grammatical and Explanatory Notes and Copious Vocabularies.
2. CÆSAR'S COMMENTARIES ON THE GALLIC WAR; with Grammatical and Explanatory Notes in English, and a complete Geographical Index.
3. CORNELIUS NEPOS; with English Notes, &c.
4. VIRGIL. THE GEORGICS, BUCOLICS, AND DOUBTFUL WORKS; with English Notes, chiefly from the German.
5. VIRGIL'S ÆNEID (on the same plan as the preceding).
6. HORACE. ODES AND EPODES; with English Notes, an Analysis of each Ode, and a full explanation of the metres.
7. HORACE. SATIRES AND EPISTLES, with English Notes, &c.
8. SALLUST. CONSPIRACY OF CATILINE, AND JUGURTHINE WAR.
9. TERENCE. ANDRIA AND HEAUTONTIMORUMENOS.
10. TERENCE. PHORMIO, ADELPHI, AND HECYRA.
11. CICERO. ORATIONS: AGAINST CATILINE, FOR SULLA, FOR ARCHIAS, AND FOR THE MANILIAN LAW.

12. CICERO. First and Second Philippics; Orations for Milo, for Marcellus, and for Ligarius.
13. CICERO. De Officiis.
14. CICERO. De Amicitiâ, de Senectute, and Brutus.
15. JUVENAL AND PERSIUS.
16. LIVY. Books I. to V., in 2 parts.
17. LIVY. Books XXI. and XXII.
18. TACITUS. Agricola; Germania; and Annals, Book I.
19. SELECTIONS FROM TIBULLUS, OVID, PROPERTIUS, AND LUCRETIUS.
20. SELECTIONS FROM SUETONIUS, AND THE LATER LATIN WRITERS.

GREEK SERIES.

1. INTRODUCTORY GREEK READER. On the same plan as the Latin Reader.
2. XENOPHON. Anabasis, I., II., III.
3. XENOPHON. Anabasis, IV., V., VI., VII.
4. LUCIAN. Select Dialogues.
5. HOMER. Iliad, I. to VI.
6. HOMER. Iliad, VII. to XII.
7. HOMER. Iliad, XIII. to XVIII.
8. HOMER. Iliad, XIX. to XXIV.
9. HOMER. Odyssey, I. to VI.
10. HOMER. Odyssey, VII. to XII.
11. HOMER. Odyssey, XIII. to XVIII.
12. HOMER. Odyssey, XIX. to XXIV.; and Hymns.
13. PLATO. Apology, Crito, and Phædo.
14. HERODOTUS, I., II.
15. HERODOTUS, III., IV.
16. HERODOTUS, V., VI., and part of VII.
17. HERODOTUS. Remainder of VII., VIII., and IX.
18. SOPHOCLES. Œdipus Rex.
19. SOPHOCLES. Œdipus Coloneus.
20. SOPHOCLES. Antigone.
21. SOPHOCLES. Ajax.
22. SOPHOCLES. Philoctetes.

23. EURIPIDES. Hecuba.
24. EURIPIDES. Medea.
25. EURIPIDES. Hippolytus.
26. EURIPIDES. Alcestis.
27. EURIPIDES. Orestes.
28. EURIPIDES. Extracts from the Remaining Plays.
29. SOPHOCLES. Extracts from the Remaining Plays.
30. ÆSCHYLUS. Prometheus Vinctus.
31. ÆSCHYLUS. Persæ.
32. ÆSCHYLUS. Septem contra Thebas.
33. ÆSCHYLUS. Choëphoræ.
34. ÆSCHYLUS. Eumenides.
35. ÆSCHYLUS. Agamemnon.
36. ÆSCHYLUS. Supplices.
37. PLUTARCH. Select Lives
38. ARISTOPHANES. Clouds.
39. ARISTOPHANES. Frogs.
40. ARISTOPHANES. Selections from the Remaining Comedies.
41. THUCYDIDES, I.
42. THUCYDIDES, II.
43. THEOCRITUS. Select Idyls.
44. PINDAR.
45. ISOCRATES.
46. HESIOD.

In one Volume large 8vo., with 13 *Plates*, *Price One Guinea*,
in *half-morocco binding*,

MATHEMATICS

FOR

PRACTICAL MEN:

BEING

A COMMON-PLACE BOOK

OF

PURE AND MIXED MATHEMATICS,

DESIGNED CHIEFLY FOR THE USE OF

CIVIL ENGINEERS, ARCHITECTS, AND SURVEYORS.

BY OLINTHUS GREGORY, LL.D., F.R.A.S

THIRD EDITION, REVISED AND ENLARGED.

BY HENRY LAW,

CIVIL ENGINEER.

CONTENTS.

PART I.—PURE MATHEMATICS.

PART II.—MIXED MATHEMATICS.

APPENDIX

In 18mo., in boards, comprising 390 pages, price 5s.

A SYNOPSIS OF PRACTICAL PHILOSOPHY,

alphabetically arranged, containing a great variety of Theorems, Formulæ, and Tables, from the most accurate and recent authorities in various branches of Mathematics and Natural Philosophy : with Tables of Logarithms.

By the Rev. JOHN CARR, M.A., late Fellow of Trinity College, Cambridge.

MR. WEALE'S
PUBLICATIONS FOR 1861.

RUDIMENTARY SERIES.

In demy 12mo, cloth, price 1s.

RUDIMENTARY.—1.—CHEMISTRY, by Professor FOWNES, F.R.S., including Agricultural Chemistry, for the Use of Farmers.

In demy 12mo, with Woodcuts, cloth, price 1s.

RUDIMENTARY.—2.—NATURAL PHILOSOPHY, by CHARLES TOMLINSON.

In demy 12mo, with Woodcuts, cloth, price 1s. 6d.

RUDIMENTARY.—3.—GEOLOGY, by Major-Gen. PORTLOCK, F.R.S., &c.

In demy 12mo, with Woodcuts, cloth, price 2s.

RUDIMENTARY.—4, 5.—MINERALOGY, with Mr. DANA'S Additions. 2 vols. in 1.

In demy 12mo, with Woodcuts, cloth, price 1s.

RUDIMENTARY.—6.—MECHANICS, by CHARLES TOMLINSON.

In demy 12mo, with Woodcuts, cloth, price 1s. 6d.

RUDIMENTARY.—7.—ELECTRICITY, by Sir WILLIAM SNOW HARRIS, F.R.S.

In demy 12mo, with Woodcuts, cloth, price 1s. 6d.

RUDIMENTARY.—7*.—ON GALVANISM; ANIMAL AND VOLTAIC ELECTRICITY; by Sir W. SNOW HARRIS.

In demy 12mo, with Woodcuts, cloth, price 3s. 6d.

RUDIMENTARY.—8, 9, 10—MAGNETISM, Concise Exposition of, by Sir W. SNOW HARRIS, 3 vols. in 1.

In demy 12mo, with Woodcuts, cloth, price 2s.

RUDIMENTARY.—11, 11*.—ELECTRIC TELEGRAPH, History of the, by E. HIGHTON, C.E.

In demy 12mo, with Woodcuts, cloth, price 1s.

RUDIMENTARY.—12.—PNEUMATICS, by CHARLES TOMLINSON.

In demy 12mo, with Woodcuts, cloth, price 4s. 6d.

RUDIMENTARY.—13, 14, 15, 15*.—CIVIL ENGINEERING, by HENRY LAW, C.E., 3 vols.; and Supplement by G. R. BURNELL, C.E.

In demy 12mo, with Woodcuts, cloth, price 1s.

RUDIMENTARY.—16.—ARCHITECTURE, Orders of, by W. H. LEEDS.

In demy 12mo, with Woodcuts, cloth, price 1s. 6d.

RUDIMENTARY.—17.—ARCHITECTURE Styles of, by T. BURY, Architect.

John Weale, 59, High Holborn, London, W.C.

B

MR. WEALE'S RUDIMENTARY SERIES.

In demy 12mo, with Woodcuts, cloth, price 2s.

RUDIMENTARY.—18, 19.—ARCHITECTURE, Principles of Design in by E. L. GARBETT, Architect, 2 vols. in 1.

In demy 12mo, with Woodcuts, cloth, price 2s.

RUDIMENTARY. — 20, 21. — PERSPECTIVE, by G. PYNE, Artist, 2 vols. in 1.

In demy 12mo, with Woodcuts, cloth, price 1s.

RUDIMENTARY.—22.—BUILDING, Art of, by E. DOBSON, C.E.

In demy 12mo, with Woodcuts, cloth, price 2s.

RUDIMENTARY.—23, 24.—BRICK-MAKING, TILE-MAKING, &c., Art of, by E. DOBSON, C.E., 2 vols. in 1.

In demy 12mo, with Woodcuts, cloth, price 2s.

RUDIMENTARY.—25, 26.—MASONRY AND STONE-CUTTING, Art of, by E. DOBSON, C.E., 2 vols. in 1.

In demy 12mo, with Woodcuts, cloth, price 2s.

RUDIMENTARY.—27, 28.—PAINTING, Art of, or a GRAMMAR OF COLOURING, by GEORGE FIELD, 2 vols. in 1.

In demy 12mo, with Woodcuts, cloth, price 1s.

RUDIMENTARY.—29.—PRACTICE OF DRAINING DISTRICTS AND LANDS, Art of, by G. D. DEMPSEY, C.E.

In demy 12mo, with Woodcuts, cloth, price 1s. 6d.

RUDIMENTARY.—30.—PRACTICE OF DRAINING AND SEWAGE OF TOWNS AND BUILD-INGS, Art of, by G. D. DEMPSEY, C.E.

In demy 12mo, with Woodcuts, cloth, price 1s.

RUDIMENTARY. — 31. — WELL-SINKING AND BORING, Art of, by G. R. BURNELL, C.E.

In demy 12mo, with Woodcuts, cloth, price 1s.

RUDIMENTARY. — 32. — USE OF INSTRU-MENTS, Art of the, by J. F. HEATHER, M.A.

In demy 12mo, with Woodcuts, cloth, price 1s.

RUDIMENTARY. — 33. — CONSTRUCTING CRANES, Art of, by J. GLYNN, F.R.S., C.E.

In demy 12mo, with Woodcuts, cloth, price 1s.

RUDIMENTARY. — 34. — STEAM ENGINE, Treatise on the, by Dr. LARDNER.

In demy 12mo, with Woodcuts, cloth, price 1s.

RUDIMENTARY.—35. — BLASTING ROCKS AND QUARRYING, AND ON STONE, by Lieut.-Gen. Sir J BURGOYNE, Hart., G.C.B., R.E.

In demy 12mo, with Woodcuts, cloth, price 4s.

RUDIMENTARY.—36, 37, 38, 39.—DICTION-ARY OF TERMS used by Architects, Builders, Civil and Mechanical Engineers, Surveyors, Artists, Ship-builders, &c., vols. in 1.

In demy 12mo, cloth, price 1s.

RUDIMENTARY.—40.—GLASS STAINING Art of, by Dr. M. A. GESSERT.

John Weale, 59, High Holborn, London, W.C.

MR. WEALE'S RUDIMENTARY SERIES.

In demy 12mo, cloth, price 1s.

RUDIMENTARY. — 41. — PAINTING ON GLASS, Essay on, by E. O. FROMBERG.

In demy 12mo, with Woodcuts. cloth, price 1s.

RUDIMENTARY.—42.— COTTAGE BUILD-ING, Treatise on.

In demy 12mo, with Woodcuts, cloth, price 1s.

RUDIMENTARY. — 43. — TUBULAR AND GIRDER BRIDGES, and others, Treatise on, more particularly describing the Britannia and Conway Bridges.

In demy 12mo, with Woodcuts, cloth, price 1s.

RUDIMENTARY.—44.—FOUNDATIONS, &c., by E. DOBSON, C.E.

In demy 12mo, with Woodcuts, cloth, price 1s.

RUDIMENTARY. — 45. — LIMES, CEMENTS, MORTARS, CONCRETE, MASTICS, &c., by G. R. BURNELL, C.E.

In demy 12mo, with Woodcuts, cloth, price 1s.

RUDIMENTARY. — 46. — CONSTRUCTING AND REPAIRING COMMON ROADS, by H. LAW, C.E.

In demy 12mo, with Woodcuts, cloth, price 3s.

RUDIMENTARY. — 47, 48, 49. — CONSTRUCTION AND ILLUMINATION OF LIGHTHOUSES, by ALAN STEVENSON. C.E., 3 vols. in 1.

In demy 12mo, with Woodcuts, cloth, price 1s.

RUDIMENTARY.—50.—LAW OF CONTRACTS FOR WORKS AND SERVICES, by DAVID GIBBONS, S.P.

In demy 12mo, with Woodcuts, cloth, price 3s.

RUDIMENTARY.—51, 52, 53.—NAVAL ARCHITECTURE, Principles of the Science, by J. PEAKE, N.A., 3 vols. in 1.

In demy 12mo, with Woodcuts, cloth, price 1s.

RUDIMENTARY AND ELEMENTARY.—53*. —PRACTICAL CONSTRUCTION concisely stated of Ships for Ocean or River Service, by Captain H. A. SOMMERFELDT, N.R.N.

In royal 4to with Engraved Plates, cloth, price 7s. 6d.

RUDIMENTARY.—53**—ATLAS of 15 Plates to ditto, drawn and engraved to a Scale for Practice.—For the convenience of the Operative Ship Builder the Atlas may be had in three separate Parts. Part I., 2s 6d. Part II., 2s. 6d. Part III., 2s. 6d.

In demy 12mo, with Woodcuts, cloth, price 1s. 6d.

RUDIMENTARY. — 54. — MASTING, MASTMAKING, AND RIGGING OF SHIPS, by R. KIPPING, A.

In demy 12mo, with Woodcuts, cloth, price 2s. 6d.

RUDIMENTARY.—54*.—IRON SHIP BUILDING, by JOHN GRANTHAM, N.A. and C.E.

In demy 12mo, with Woodcuts, cloth, price 2s.

RUDIMENTARY. — 55, 56.— NAVIGATION ; THE SAILOR'S SEA-BOOK.—How to Keep the Log and Work it Off—Latitude and Longitude—Great Circle Sailing—Law of Storms and variable Winds; and an Explanation of Terms used, with coloured Illustrations of Flags.

John Weale, 59, High Holborn, London, W.C.

MR. WEALE'S RUDIMENTARY SERIES.

In demy 12mo, with Woodcuts, cloth, price 2s.

RUDIMENTARY.—57, 58.—WARMING AND VENTILATION, by CHARLES TOMLINSON, 2 vols. in 1.

In demy 12mo, with Woodcuts, cloth, price 1s.

RUDIMENTARY.—59.—STEAM BOILERS, by R. ARMSTRONG, C.E.

In demy 12mo, with Woodcuts, cloth, price 2s.

RUDIMENTARY. — 60, 61. — LAND AND ENGINEERING SURVEYING, by T. BAKER, C.E., 2 vols. in 1.

In demy 12mo, with Woodcuts, cloth, price 1s.

RUDIMENTARY AND ELEMENTARY.—62. —PRINCIPLES OF RAILWAYS, for the Use of the Beginner in his Studies; with Sketches for Construction. By Sir R. MACDONALD STEPHENSON. Vol. I.

In demy 12mo, with Woodcuts, cloth, price 1s.

RUDIMENTARY.—62*.—RAILWAY WORKING IN GREAT BRITAIN, Statistical Details, Table of Capital and Dividends, Revenue Accounts, Signals, &c., Vol. II.

In demy 12mo, with Woodcuts, cloth, price 3s.

RUDIMENTARY.—63, 64, 65.—AGRICULTURAL BUILDINGS, the Construction of, on Motive Powers, and the Machinery of the Steading; and on Agricultural Field-Engines, Machines, and Implements, by G. H. ANDREWS, 3 vols in 1. –John Weale, 59, High Holborn, London, W.C.

In demy 12mo, cloth, price 1s.

RUDIMENTARY.—66.—CLAY LANDS AND LOAMY SOILS, by Professor JOHN DONALDSON, A.E.

In demy 12mo, with Woodcuts, cloth, price 3s.

RUDIMENTARY. — 67, 68. — CLOCK AND WATCH-MAKING, AND ON CHURCH CLOCKS AND BELLS, by E. B. DENISON, M.A., 2 vols. in 1, considerably extended. Fourth Edition.

In demy 12mo, with Woodcuts, cloth, price 2s.

RUDIMENTARY.—69, 70.— MUSIC, Practical Treatise on, by C. C. SPENCER, Mus. Dr. 2 vols. in 1.

In demy 12mo, cloth, price 1s.

RUDIMENTARY. — 71. — PIANOFORTE, Instruction for Playing the, by C. C. SPENCER, Mus. Dr.

In demy 12mo, with Steel Engravings and Woodcuts, cloth, price 5s. 6d.

RUDIMENTARY.—72, 73, 74, 75, 75*.—RECENT FOSSIL SHELLS (A Manual of the Mollusca), by SAMUEL P. WOODWARD, of the Brit. Mus. 4 vols. in 1, with Supplement.

In demy 12mo., with Woodcuts, cloth, price 2s.

RUDIMENTARY. — 76, 77. — DESCRIPTIVE GEOMETRY, by J. F. HEATHER, M.A. 2 vols. in 1.

In demy 12mo, with Woodcuts, price 1s.

RUDIMENTARY. — 77*. — ECONOMY OF FUEL, by T. S. PRIDEAUX.

In demy 12mo, 2 vols. in 1, with Woodcuts, cloth, price 2s.

RUDIMENTARY.—78, 79.—STEAM AS APPLIED TO GENERAL PURPOSES.
John Weale, 59, High Holborn, London, W.C.

MR. WEALE'S RUDIMENTARY SERIES.

In demy 12mo, with Woodcuts, cloth, price 1s. 6d.
RUDIMENTARY.—78*.—LOCOMOTIVE EN-GINE, by G. D. DEMPSEY. C.E.

In royal 4to, cloth, price 4s. 6d.
RUDIMENTARY. — 79*. — ATLAS OF EN-GRAVED PLATES to DEMPSEY'S LOCOMOTIVE ENGINES.

In demy 12mo, with Woodcuts, cloth, price 1s.
RUDIMENTARY.—79**.—ON PHOTOGRA-PHY, the Composition and Properties of the Chemical Sub-stances used, by Dr. H. HALLEUR.

In demy 12mo., with Woodcuts, cloth, price 2s. 6d.
RUDIMENTARY. — 80, 81. — MARINE EN-GINES AND ON THE SCREW, &c., by R. MURRAY, C.E. 2 vols. in 1.

In demy 12mo, cloth, price 2s.
RUDIMENTARY.—80*, 81*.—EMBANKING LANDS FROM THE SEA, by JOHN WIGGINS, F.G.S. 2 vols. in 1.

In demy 12mo, with Woodcuts, cloth, price 2s.
RUDIMENTARY. — 82, 82*. — POWER OF WATER, AS APPLIED TO DRIVE FLOUR MILLS, by JOSEPH GLYNN, F.R.S., C.E.

In demy 12mo, cloth, price 1s.
RUDIMENTARY.—83.—BOOK-KEEPING, by JAMES HADDON, M.A.

In demy 12mo, with Woodcuts, price 3s.
RUDIMENTARY. — 82**, 83*, 83 (bis) COAL GAS, on the Manufacture and Distribution of, by SAMUEL HUGHES, C.E.

In demy 12mo, with Woodcuts, cloth, price 3s.
RUDIMENTARY.—82***.—WATER WORKS FOR THE SUPPLY OF CITIES AND TOWNS; Works which have been executed for procuring Supplies by means of Drainage Areas and by Pumping from Wells, by SAMUEL HUGHES, C.E.

In demy 12mo, with Woodcuts, cloth, price 1s. 6d.
RUDIMENTARY. —83**.— CONSTRUCTION OF DOOR LOCKS.

In demy 12mo, with Woodcuts, cloth, price 1s.
RUDIMENTARY. — 83 (bis) — FORMS OF SHIPS AND BOATS, by W. BLAND, of Hartlip.

In demy 12mo, cloth, price 1s. 6d.
RUDIMENTARY.—84.—ARITHMETIC, with numerous Examples, by Prof. J. R. YOUNG.

In demy 12mo, cloth, price 1s. 6d.
RUDIMENTARY. —84*.— KEY to the above, by Prof. J. R. YOUNG.

In demy 12mo, cloth, price 1s.
RUDIMENTARY. — 85. — EQUATIONAL ARITHMETIC, Questions of Interest, Annuities, &c., by W. HIPSLEY.

John Weale, 59, High Holborn, London, W.C.

6

In demy 12mo, cloth, price 1s.

RUDIMENTARY.—85*.—SUPPLEMENTARY VOLUME TO HIPSLEY'S EQUATIONAL ARITHME-TIC, Tables for the Calculation of Simple Interest, with Logarithms for Compound Interest and Annuities, &c , &c., by W. HIPSLEY.

In demy 12mo, cloth, price 2s.

RUDIMENTARY. — 86, 87. — ALGEBRA, by JAMES HADDON, M.A. 2 vols. in 1.

In demy 12mo, in cloth, price 1s. 6d.

RUDIMENTARY.—86*, 87*.—ELEMENTS OF ALGEBRA, Key to the, by Prof. YOUNG.

In demy 12mo, with Woodcuts, price 2s.

RUDIMENTARY.—88, 89.—ELEMENTS OF GEOMETRY, by HENRY LAW, C.E. 2 vols. in 1.

In demy 12mo, with Woodcuts, cloth, price 1s.

RUDIMENTARY.—90.—GEOMETRY, ANA-LYTICAL, by Prof. JAMES HANN.

In demy 12mo, with Woodcuts, cloth, price 2s.

RUDIMENTARY. — 91, 92. — PLANE AND SPHERICAL TRIGONOMETRY, by the same. 2 vols. in 1.

In demy 12mo, with Woodcuts, cloth, price 1s.

RUDIMENTARY.—93.—MENSURATION, by T. BAKER, C.E.

In demy 12mo, cloth, price 2s. 6d,

RUDIMENTARY. — 94, 95. — LOGARITHMS, Tables for facilitating Astronomical, Nautical, Trigonometri-cal, and Logarithmic Calculations, by H. LAW, C.E. New Edition, with Tables of Natural Sines and Tangents, and Natural Cosines. 2 vols. in 1.

In demy 12mo, with Woodcuts, cloth, price 1s.

RUDIMENTARY.—96.—POPULAR ASTRO-NOMY. By the Rev. ROBERT MAIN, M.R.A.S.

In demy 12mo, with Woodcuts, cloth, price 1s.

RUDIMENTARY.—97.—STATICS AND DY-NAMICS, by T. BAKER, C.E.

In demy 12mo, with 2*) Woodcuts, cloth, price 2s. 6d.

RUDIMENTARY. — 98, 98*. — MECHANISM AND PRACTICAL CONSTRUCTION OF MACHINES, by T. BAKER, C.E., and ON TOOLS AND MACHINES, by JAMES NASMYTH, C.E.

In demy 12mo, with Woodcuts, cloth, price 2s.

RUDIMENTARY.—99, 100.—NAUTICAL AS-TRONOMY AND NAVIGATION, by Prof. YOUNG. 2 vols. in 1.

In demy 12mo, cloth, price 1s. 6d.

RUDIMENTARY. — 100*. — NAVIGATION TABLES, compiled for practical use with the above.

In demy 12mo, cloth, price 1s.

RUDIMENTARY. — 101. — DIFFERENTIAL CALCULUS, by Mr. WOOLHOUSE, F.R.A.S. John Weale, 59, High Holborn, London, W.C.

7

MR. WEALE'S RUDIMENTARY SERIES.

In demy 12mo, cloth, price 1s. 6d.
RUDIMENTARY. — 101*. — WEIGHTS AND
MEASURES OF ALL NATIONS: Weights, Coins, and the
various Divisions of Time, with the principles which determine
Rates of Exchange, by Mr. WOOLHOUSE, F.R.A.S.

In demy 12mo, in cloth, price 1s.
RUDIMENTARY. — 102. — INTEGRAL CAL-
CULUS, by H. COX, M.A.

In demy 12mo, in cloth, price 1s.
RUDIMENTARY. — 103. — INTEGRAL CAL-
CULUS, Examples of, by Prof. JAMES HANN.

In demy 12mo, cloth, price 1s.
RUDIMENTARY. — 104. — DIFFERENTIAL
CALCULUS, Examples of, by J. HADDON, M A.

In demy 12mo, with Woodcuts, cloth, price 1s. 6d.
RUDIMENTARY. — 105. — ALGEBRA, GEO-
METRY, AND TRIGONOMETRY, Mnemonical Lessons,
by the Rev. T. PENYNGTON KIRKMAN, M.A.

In demy 12mo, with Woodcuts, cloth, price 1s. 6d.
RUDIMENTARY. — 106. — SHIPS' ANCHORS
FOR ALL SERVICES, by Mr. GEORGE COTSELL, N.A.

In demy 12mo, with Woodcuts, price 2s. 6d.
RUDIMENTARY. — 107. — METROPOLITAN
BUILDINGS ACT in present operation, with Notes, and the
Act dated August 29th, 1860, for better supplying of Gas to the
Metropolis.

In demy 12mo, cloth, price 1s. 6d.
RUDIMENTARY. — 108. — METROPOLITAN
LOCAL MANAGEMENT ACTS. All the Acts.

In demy 12mo, cloth, price 1s. 6d.
RUDIMENTARY, — 109. — LIMITED LIA-
BILITY AND PARTNERSHIP ACTS.

In demy 12mo, cloth, price 1s.
RUDIMENTARY. — 110. — SIX RECENT LE-
GISLATIVE ENACTMENTS, for Contractors, Merchants,
and Tradesmen.

In demy 12mo, cloth, price 1s.
RUDIMENTARY. — 111. — NUISANCES RE-
MOVAL AND DISEASE PREVENTION ACT.

In demy 12mo, cloth, price 1s 6d.
RUDIMENTARY. — 112. — DOMESTIC MEDI-
CINE, PRESERVING HEALTH, by M. RASPAIL.

In demy 12mo, cloth, price 1s. 6d.
RUDIMENTARY. — 113. — USE OF FIELD
ARTILLERY ON SERVICE, by Lieut.-Col. HAMILTON
MAXWELL, B.A.

In demy 12mo, with Woodcuts, cloth, price 1s. 6d.
RUDIMENTARY. — 114. — ON MACHINERY:
Rudimentary and Elementary Principles of the Construction
and on the Working of Machinery, by C. D. ABEL, C.E.

In royal 4to, cloth, price 7s. 6d.
RUDIMENTARY. — 115. — ATLAS OF PLATES
OF SEVERAL KINDS OF MACHINES, 17 very valuable
Illustrative plates.

John Weale, 59, High Holborn, London, W.C.

8

MR. WEALE'S RUDIMENTARY SERIES.

In demy 12mo, with Woodcuts, cloth, price 1s. 6d.

RUDIMENTARY. — 116. — TREATISE ON ACOUSTICS: The Distribution of Sound, by T. ROGER SMITH, Architect.

In demy 12mo, with Woodcuts, cloth, price 2s. 6d.

RUDIMENTARY.—117.—SUBTERRANEOUS SURVEYING, RANGING THE LINE WITHOUT THE MAGNET. By THOMAS FENWICK, Coal Viewer. With Improvements and Modern Additions by T. BAKER, C.E.

In demy 12mo, with Plates and Woodcuts, cloth, price 3s.

RUDIMENTARY.—118, 119.—ON THE CIVIL ENGINEERING OF NORTH AMERICA, by D. STEVENSON, C.E. 2 vols. in 1.

In demy 12mo, with Woodcuts, cloth, price 3s.

RUDIMENTARY. — 120. — ON HYDRAULIC ENGINEERING, by G. R. BURNELL, C.E. 3 vols. in 1.

In demy 12mo, with 2 Engraved Plates, cloth, price 1s. 6d.

RUDIMENTARY. — 121. — TREATISE ON RIVERS AND TORRENTS, from the Italian of PAUL FRISI.

In demy 12mo, by PAUL FRISI, in cloth, price 1s.

RUDIMENTARY.—122.—ON RIVERS THAT CARRY SAND AND MUD, and an ESSAY ON NAVIGABLE CANALS. 121 and 122 bound together, 2s. 6d.

In demy 12mo, with Woodcuts, cloth, price 1s. 6d.

RUDIMENTARY. — 123. — ON CARPENTRY AND JOINERY, founded on Dr. Robson's Work.

In demy 4to, cloth, price 4s. 6d.

RUDIMENTARY.—123*.—ATLAS of PLATES in detail to the CARPENTRY AND JOINERY. 123 and 123* bound together in cloth in 1 vol.

In demy 12mo, with Woodcuts, cloth, price 1s. 6d.

RUDIMENTARY. — 124. — ON ROOFS FOR PUBLIC AND PRIVATE BUILDINGS, founded on Dr. Robison's Work.

In royal 4to, cloth, price 4s. 6d.

RUDIMENTARY.—124*.—RECENTLY CONSTRUCTED IRON ROOFS, Atlas of plates.

In demy 12mo, with Woodcuts, cloth, price 3s.

RUDIMENTARY.— 125.—ON THE COMBUSTION OF COAL AND THE PREVENTION OF SMOKE, Chemically and Practically Considered, by CHARLES WYE WILLIAMS.

In demy 12mo, cloth. 125 and 126 together, price 3s.

RUDIMENTARY. — 126. — ILLUSTRATIONS to WILLIAMS'S COMBUSTION OF COAL. 125 and 126, 2 vols. bound in 1.

In demy 12mo, with Woodcuts, cloth, price 1s. 6d.

RUDIMENTARY. — 127. — PRACTICAL INSTRUCTIONS IN THE ART OF ARCHITECTURAL MODELLING.

John Weale, 59, High Holborn, London, W.C.

MR. WEALE'S RUDIMENTARY SERIES.

In demy 12mo, with Engravings and Woodcuts.

RUDIMENTARY.—128.—THE TEN BOOKS OF M. VITRUVIUS ON CIVIL, MILITARY, AND NAVAL ARCHITECTURE, translated by JOSEPH GWILT, Arch. 2 vols. in 1.

In demy 12mo, 128 and 129 together, cloth, price 5s.

RUDIMENTARY. — 129. — ILLUSTRATIVE PLATES TO VITRUVIUS'S TEN BOOKS, by the Author and JOSEPH GANDY, R.A.

In demy 12mo. cloth, price 1s.

RUDIMENTARY.—130.— INQUIRY INTO THE PRINCIPLES OF BEAUTY IN GRECIAN ARCHITECTURE, by the Right Hon. the Earl of ABERDEEN, &c. &c.

In demy 12mo, cloth, price 1s.

RUDIMENTARY. — 131. — THE MILLER'S, MERCHANT'S, AND FARMER'S READY RECKONER, for ascertaining at Sight the Value of any quantity of Corn; together with the approximate value of Millstones and Millwork.

In demy 12mo, with Woodcuts, cloth, price 2s. 6d.

RUDIMENTARY.—132.—TREATISE ON THE ERECTION OF DWELLING HOUSES. WITH SPECIFICATIONS, QUANTITIES OF THE VARIOUS MATERIALS, &c., by S. H. BROOKS, Architect. 27 Plates.

RUDIMENTARY SERIES. — ON MINES, SMELTING WORKS, AND THE MANUFACTURE OF METALS, as follows.

In demy 12mo, with Woodcuts, cloth, price 2s.

RUDIMENTARY. — Vol. 1. — TREATISE ON THE METALLURGY OF COPPER, by R. H. LAMBORN.

In demy 12mo, to have Woodcuts, cloth.

RUDIMENTARY. — Vol. 2. — TREATISE ON THE METALLURGY OF SILVER AND LEAD.

In demy 12mo, to have Woodcuts, cloth.

RUDIMENTARY AND ELEMENTARY.— Vol. 3.—TREATISE ON IRON METALLURGY up to the Manufacture of the latest processes.

In demy 12mo, to have Woodcuts, cloth.

RUDIMENTARY AND ELEMENTARY. — Vol. 4.—TREATISE ON GOLD MINING AND ASSAYING PLATINUM, IRIDIUM, &c.

In demy 12mo, to have Woodcuts, cloth.

RUDIMENTARY AND ELEMENTARY. — Vol. 5.—TREATISE ON THE MINING OF ZINC, TIN, NICKEL, COBALT, &c.

In demy 12mo, to have Woodcuts, cloth.

RUDIMENTARY AND ELEMENTARY. — Vol. 6.—TREATISE ON COAL MINING (Geology and Means of Discovering, &c.)

In demy 12mo, with Woodcuts, cloth, price 1s. 6d.

RUDIMENTARY. — Vol. 7. — ELECTRO-METALLURGY. — Practically treated by ALEXANDER WATT, F.R.S.A.

John Weale, 59, High Holborn, London, W.C.

B 2

NEW SERIES OF EDUCATIONAL WORKS.

In demy 12mo, with Woodcuts, cloth, price 4s.

CONSTITUTIONAL HISTORY OF ENG-
LAND.—1, 2, 3, 4.—By W. D. HAMILTON, of the State P. O.

In demy 12mo, with Woodcuts, cloth, price 2s. 6d.

OUTLINES OF THE HISTORY OF GREECE.
—5, 6.—By W. D. HAMILTON, 2 vols.

In demy 12mo, with Map of Italy and Woodcuts, cloth, price 2s. 6d

OUTLINE OF THE HISTORY OF ROME.—
7, 8.—By W. D. HAMILTON, 2 vols.

In demy 12mo, cloth, price 2s. 6d.

CHRONOLOGY OF CIVIL AND ECCLESI-
ASTICAL HISTORY, LITERATURE, ART, AND CIVI-
LISATION, from the earliest period to the present.—9, 10.—2 vols.

In demy 12mo, cloth, price 1s.

GRAMMAR OF THE ENGLISH LANGUAGE.
—11.—By HYDE CLARKE, D.C.L.

In demy 12mo, cloth, price 1s.

HANDBOOK OF COMPARATIVE PHILO-
LOGY.—11*.—By HYDE CLARKE, D.C.L.

In demy stout 12mo, cloth, price 3s. 6d.

DICTIONARY OF THE ENGLISH LAN-
GUAGE.—12, 13.—A New Dictionary of the English Tongue
as spoken and written, above 100,000 words, or 50,000 more than in
any existing work, by HYDE CLARKE, D.C.L., 3 vols. in 1.

In demy 12mo, cloth, price 1s.

GRAMMAR OF THE GREEK LANGUAGE.
—14.—By H. C. HAMILTON.

In demy 12mo, cloth, price 2s.

DICTIONARY OF THE GREEK AND ENG-
LISH LANGUAGES.—15, 16.—By H. R. HAMILTON, 2
vols. in 1.

In demy 12mo, cloth, price 2s.

DICTIONARY OF THE ENGLISH AND
GREEK LANGUAGES.—17, 18.—By H. R. HAMILTON, 2
vols. in 1.

In demy 12mo, cloth, price 1s.

GRAMMAR OF THE LATIN LANGUAGE.
—19.—By the Rev. T. GOODWIN, A.B.

In demy 12mo, cloth, price 2s.

DICTIONARY OF THE LATIN AND ENG-
LISH LANGUAGES.—20, 21.—By the Rev. T. GOODWIN,
B.A. Vol. 1.

In demy 12mo, cloth, price 1s. 6d.

DICTIONARY OF THE ENGLISH AND
LATIN LANGUAGES.—22, 23.—By the Rev. T. GOOD-
WIN, A.B. Vol. II.

In demy 12mo, cloth, price 1s.

GRAMMAR OF THE FRENCH LANGUAGE.
—24.

John Weale, 59, High Holborn, London, W.C.

MR. WEALE'S EDUCATIONAL SERIES.

In demy 12mo, cloth, price 1s.

DICTIONARY OF THE FRENCH AND ENGLISH LANGUAGES.—25.—By A. ELWES. Vol. I.

In demy 12mo, cloth, price 1s. 6d.

DICTIONARY OF THE ENGLISH AND FRENCH LANGUAGES.—26.—By A. ELWES. Vol. II.

In demy 12mo, cloth, price 1s.

GRAMMAR OF THE ITALIAN LANGUAGE —27.—By A. ELWES.

In demy 12mo, cloth, price 2s.

DICTIONARY OF THE ITALIAN, ENGLISH, AND FRENCH LANGUAGES.—28, 29.—By A. ELWES. Vol. I.

In demy 12mo, cloth, price 2s.

DICTIONARY OF THE ENGLISH, ITALIAN, AND FRENCH LANGUAGES.—30, 31.—By A. ELWES. Vol. II.

In demy 12mo, cloth, price 2s.

DICTIONARY OF THE FRENCH, ITALIAN, AND ENGLISH LANGUAGES.—32, 33.—By A. ELWES. Vol. III.

In demy 12mo, cloth, price 1s.

GRAMMAR OF THE SPANISH LANGUAGE. —34.—By A. ELWES.

In demy 12mo, cloth, price 4s.

DICTIONARY OF THE SPANISH AND ENGLISH LANGUAGES.—35, 36, 37, 38.—By A. ELWES. 4 vols. in 1.

In demy 12mo, cloth, price 1s.

GRAMMAR OF THE GERMAN LANGUAGE. —39.

In demy 12mo, cloth, price 1s.

CLASSICAL GERMAN READER.—40.—From the best Authors.

In demy 12mo, cloth, price 3s.

DICTIONARIES OF THE ENGLISH, GERMAN, AND FRENCH LANGUAGES.—41, 42, 43.—By N. E. HAMILTON, 3 vols., separately, 1s. each.

In demy 12mo, cloth, price 7s.

DICTIONARY OF THE HEBREW AND ENGLISH LANGUAGES.—44, 45.—Containing the Biblical and Rabbinical words, 2 vols. (together with the Grammar, which may be had separately for 1s.), by Dr. BRESSLAU, Hebrew Professor.

In demy 12mo, cloth, price 3s.

DICTIONARY OF THE ENGLISH AND HEBREW LANGUAGES.—46.—Vol. III. to complete.

In demy 12mo, cloth, price 1s.

FRENCH AND ENGLISH PHRASE BOOK. —47.

John Weale, 59, High Holborn, London, W.C.

MR. WEALE'S CLASSICAL SERIES.

Now in course of Publication, in demy 12mo, price 1s. per Volume (except in some instances, and those are 1s. 6d. or 2s. each), very neatly printed on good paper. Those priced are published.

GREEK AND LATIN CLASSICS.—A Series of Volumes containing the principal Greek and Latin Authors, accompanied by Explanatory Notes in English, principally selected from the best and most recent German Commentators, and comprising all those Works that are essential for the Scholar and the Pupil, and applicable for the Universities of Oxford, Cambridge, Edinburgh, Glasgow, Aberdeen, and Dublin—the Colleges at Belfast, Cork, Galway, Winchester, and Eton, and the great Schools at Harrow, Rugby, &c —also for Private Tuition and Instruction, and for the Library, as follows:

LATIN SERIES.

In demy 12mo, boards, price 1s.

A NEW LATIN DELECTUS.—1.—Extracts from Classical Authors, with Vocabularies and Explanatory Notes.

In demy 12mo, boards, price 2s.

CÆSAR'S COMMENTARIES ON THE GALLIC WAR.—2.—With Grammatical and Explanatory Notes in English, and a Geographical Index.

In demy 12mo, boards, price 1s.

CORNELIUS NEPOS.—3.—With English Notes, &c.

In demy 12mo, boards, price 1s.

VIRGIL—4.—The Georgics, Bucolics, with English Notes.

In demy 12mo, boards, price 2s.

VIRGIL'S ÆNEID.—5.—(On the same plan as the preceding).

In demy 12mo, boards, price 1s.

HORACE.—6.—Odes and Epodes ; with English Notes, and Analysis and Explanation of the Metres.

In demy 12mo, boards, price 1s. 6d.

HORACE.—7.—Satires and Epistles, with English Notes, &c.

In demy 12mo, boards, price 1s. 6d.

SALLUST.—8.—Conspiracy of Catiline, Jugurthine War, with English Notes.

In demy 12mo, boards, price 1s. 6d.

TERENCE.—9.—Andrea and Heautontimorumenos, with English Notes.

In demy 12mo, boards, price 2s.

TERENCE.—10.—Phormio, Adelphi, and Hecyra, with English Notes.

In demy 12mo.

CICERO.—11.—Orations against Catiline, for Sulla, for Archias, and for the Manilian Law.

In demy 12mo.

CICERO.—12.—First and Second Philippics ; Orations for Milo, for Marcellus, &c.

John Weale, 59, High Holborn London, W.C.

13

MR. WEALE'S CLASSICAL SERIES.

In demy 12mo.
CICERO.—13.—De Officiis.

In demy 12mo, boards, price 2s.
CICERO.—14.—De Amicitiâ, de Senectute, and Brutus, with English Notes.

In demy 12mo.
JUVENAL AND PERSIUS.—15.—(The indelicate parts expunged.)

In demy 12mo, boards, price 3s.
LIVY. — 16. — Books i. to v. in two vols., with English Notes.

In demy 12mo, boards, price 1s.
LIVY.—17.—Books xxi. and xxii., with English Notes.

In demy 12mo.
TACITUS.—18.—Agricola; Germania; and Annals, Book i.

In demy 12mo, boards, price 2s.
SELECTIONS FROM TIBULLUS, OVID, and PROPERTIUS.—19.—With English Notes,

In demy 12mo.
SELECTIONS FROM SUETONIUS and the later Latin Writers.—20.

GREEK SERIES, ON A SIMILAR PLAN TO THE LATIN SERIES.
Those not priced are in the Press.

In demy 12mo, boards, price 1s.
INTRODUCTORY GREEK READER. — 1. — On the same plan as the Latin Reader.

In demy 12mo, boards, price 1s.
XENOPHON. — 2. — Anabasis, i. ii. iii., with English Notes.

In demy 12mo, boards, price 1s.
XENOPHON. — 3. — Anabasis, iv. v. vi. vii., with English Notes.

In demy 12mo, boards, price 1s.
LUCIAN. — 4. — Select Dialogues, with English Notes.

In demy 12mo, boards, price 1s. 6d.
HOMER.—5.—Iliad, i. to vi., with English Notes.

In demy 12mo, boards, price 1s. 6d.
HOMER.—6.—Iliad, vii. to xii., with English Notes.

In demy 12mo, boards, price 1s. 6d.
HOMER. — 7. — Iliad, xiii. to xviii. with English Notes.

In demy 12mo, boards, price 1s. 6d.
HOMER. — 8. — Iliad, xix. to xxiv., with English Notes.

John Weale, 59, High Holborn, London, W.C.

MR. WEALE'S CLASSICAL SERIES.

In demy 12mo, boards, price 1s. 6d.

HOMER.—9.—Odyssey, i. to vi., with English Notes.

In demy 12mo, boards, price 1s. 6d.

HOMER.—10.—Odyssey, vii. to xii., with English Notes.

In demy 12mo, boards, price 1s. 6d.

HOMER.—11.—Odyssey, xiii. to xviii. with English Notes.

In demy 12mo, boards, price 1s. 6d.

HOMER. — 12. — Odyssey, xix. to xxiv.; and Hymns, with English Notes.

In demy 12mo, boards, price 2s.

PLATO. — 13. — Apology, Crito, and Phædo, with English Notes.

In demy 12mo, boards, price 1s. 6d.

HERODOTUS.—14.—i. ii., with English Notes.— Dedicated to His Grace the Duke of Devonshire.

In demy 12mo, boards, price 1s. 6d.

HERODOTUS.—15.—iii. iv., with English Notes. Dedicated to His Grace the Duke of Devonshire.

In demy 12mo.

HERODOTUS.—16.—v. vi. and part of vii. Dedicated to His Grace the Duke of Devonshire.

In demy 12mo.

HERODOTUS.—17.—Remainder of vii., viii., and ix. Dedicated to His Grace the Duke of Devonshire.

In demy 12mo, boards, price 1s.

SOPHOCLES. — 18. — Œdipus Rex, with English Notes.

In demy 12mo.

SOPHOCLES.—19.—Œdipus Colonæus.

In demy 12mo.

SOPHOCLES.—20.—Antigone.

In demy 12mo.

SOPHOCLES.—21.—Ajax.

In demy 12mo.

SOPHOCLES.—22.—Philoctetes.

In demy 12mo, boards, price 1s. 6d.

EURIPIDES.—23.—Hecuba, with English Notes.

In demy 12mo.

EURIPIDES.—24.—Medea.

In demy 12mo.

EURIPIDES.—25.—Hippolytus.

John Weale, 59, High Holborn, London, W.C.

MR. WEALE'S CLASSICAL SERIES.

In demy 12mo. boards, price 1s.

E URIPIDES.—26.—Alcestis, with English Notes.

In demy 12mo.

E URIPIDES.—27.—Orestes.

In demy 12mo.

E URIPIDES.—28.—Extracts from the remaining Plays.

In demy 12mo.

S OPHOCLES.—29.—Extracts from the remaining Plays.

In demy 12mo.

Æ SCHYLUS.—30.—Prometheus Vinctus.

In demy 12mo.

Æ SCHYLUS.—31.—Persæ.

In demy 12mo.

Æ SCHYLUS.—32.—Septem contra Thebas.

In demy 12mo.

Æ SCHYLUS.—33.—Choëphoræ.

In demy 12mo.

Æ SCHYLUS.—34.—Eumenides.

In demy 12mo.

Æ SCHYLUS.—35.—Agamemnon.

In demy 12mo.

Æ SCHYLUS.—36.—Supplices.

In demy 12mo.

P LUTARCH.—37.—Select Lives.

In demy 12mo,

A RISTOPHANES.—38.—Clouds.

In demy 12mo.

A RISTOPHANES.—39.—Frogs.

In demy 12mo.

A RISTOPHANES. — 40. — Selections from the remaining Comedies.

In demy 12mo, boards, price 1s.

T HUCYDIDES. — 41. — I., with English Notes.

In demy 12mo.

T HUCYDIDES.—42.—II.

John Weale, 59, High Holborn, London, W.C.

MR. WEALE'S CLASSICAL SERIES.

In demy 12mo.

THEOCRITUS.—43.—Select Idyls.

In demy 12mo.

PINDAR.—44.

In demy 12mo.

SOCRATES.—45.

In demy 12mo.

HESIOD.—46.

MR. WEALE'S PUBLICATIONS OF WORKS ON ARCHITECTURE, ENGINEERING, AND THE FINE ARTS.

In 1 large Atlas, folio Volume, with fine Plates, price £4 4s.

"BRITISH GOVERNMENT WORK."—THE ARCHITECTURAL ANTIQUITIES AND RESTORATION OF ST. STEPHEN'S CHAPEL, WESTMINSTER (late the House of Commons).

Fine Plates and Vignettes, Atlas folio, price £3 10s.

"NORWEGIAN GOVERNMENT WORK." —THE CATHEDRAL OF THRONDHEIM, IN NORWAY. Text by Professor MUNCH; drawings by H. E. SCHIRMER, Architect.

Large Atlas folio, 4 livraisons, published in Madrid, at 100 reals each, or £1 in England. Illustrated by beautifully executed Engravings, some of which are coloured.

"SPANISH GOVERNMENT WORK."— MONUMENTS ARCHITECTONIQUES DE L'ESPAGNE, PUBLIÉS AUX FRAIS DE LA NATION.—Part I Provincia de Toledo, Granada, Alcalá de Henares.—Part 2. Catedral Toledo, Detailles.—Part 3. Granada, Segovia, Toledo, Salamanca.—Part 4. Santa Maria de Alca'á de Henares, Casa Lonia de Valencia, Toledo, Segovia, &c.—This work surpasses in beauty all other works.

Columbier folio plates, with text also uniform, with gold borders, and sumptnously bound in red morocco, gilt; gilt leaves, £12 12s., Columbier folio plates, with text also uniform, with gold borders, and elegantly half-bound in morocco, gilt, £10 10s.; Plates in Columbier folio, and text in Imperial 4to, half-bound in morocco, gilt, £7 7s.; Plates in Columbier folio, and text in Imperial 4to, in cloth extra, boards and lettered, £4 14s. 6d.

THE VICTORIA BRIDGE, AT MONTREAL, IN CANADA. — Elaborately illustrated by views, plans, elevations, and details of the Bridge; together with the Illustrations of the Machinery and Contrivances used in the construction of this stupendously important and valuable engineering work. The whole produced in the finest style of art, pictorially and geometrically drawn, and the views highly coloured, and a descriptive text. Dedicated to His Royal Highness the Prince of Wales. By JAMES HODGES, Engineer to the Contractors. Engineers: ROBERT STEPHENSON and ALEX. M. ROSS. Contractors: Sir S. MORTON PETO, Bart., M.P., THOMAS BRASSEY, and EDWARD LADD BETTS, Esqrs.

John Weale, 59, High Holborn, London, W.C.

MR. WEALE'S WORKS ON ARCHITECTURE, ENGINEERING, FINE ARTS, &c.

In 4to, 1s. 6d.

ARAGO, Mons. — Report on the Atmospheric System, and on the proposed Atmospheric Railway at Paris.

In 4to, with about 500 Engravings, some of which are highly coloured, 4 vols., original copies, half-bound in morocco, £6 6s.

ARCHITECTURAL PAPERS.

2 Engravings, in folio, useful to learners and for schools, 2s. 6d.

ARCHITECTURAL ORDERS (FIVE) AND THEIR ENTABLATURES, drawn to a larger scale, with Figured Dimensions.

4to, 1s.

ARNOLLET, M. — Report on his Atmospheric Railway.

In 4to, 10 Plates, 7s. 6d.

ATMOSPHERIC RAILWAYS. — THREE REPORTS on improved methods of Constructing and Working Atmospheric Railways. By R. MALLET, C.E.

In 8vo, 1s. 6d.

BARLOW, P. W. — Observations on the Niagara Railway Suspension Bridge.

In large 4to, very neat half-morocco, 18s., with Engravings.

BARRY, SIR CHARLES, R.A., &c. — Studies of Modern English Architecture. By W. H. LEEDS; The Travellers' Club-House, illustrated by Engravings of Plans, Sections, Elevations, and details.

In 1 Vol., large 8vo, with coloured Plates, half-morocco, price £1 1s.

BEWICK'S (J. G.) GEOLOGICAL TREATISE ON THE DISTRICT OF CLEVELAND IN NORTH YORKSHIRE, its Ferruginous Deposits, Lias and Oolites; with some Observations on Ironstone Mining.

In 8vo, with Plates. Price 4s.

BINNS, W. S. — Work on Geometrical Drawing, embracing Practical Geometry, including the use of Drawing Instruments, the construction and use of Scales, Orthographic Projection, and Elementary Descriptive Geometry.

In 4to, with 106 Illustrative Plates, cloth boards, £1 11s. 6d.

BLASHFIELD, J. M., M. R. Inst., &c.— SELECTIONS OF VASES, STATUES, BUSTS, &c, from TERRA COTTAS.

In 8vo, Woodcuts, 1s.

BLASHFIELD, J. M., M. R., Inst., &c.— ACCOUNT OF THE HISTORY AND MANUFACTURE OF ANCIENT AND MODERN TERRA COTTA.

In 4to, 2s. 6d.

BODMER, R., C.E.—On the Propulsion of Vessels by the Screw.

15s.

BRIDGE. — A large magnificent Plate, 3 feet 6 inches by 2 feet, on a scale of 25 feet to an inch, of LONDON BRIDGE; containing Plan and Elevation. Engraved and elaborately finished. The Work of the RENNIES.

John Weale, 59, High Holborn, London, W.C.

MR. WEALE'S WORKS ON ARCHITEC-
TURE, ENGINEERING, FINE ARTS, &c.

10s.

BRIDGE. — Plan and Elevation, on a scale of 10 feet to an inch, of STAINES BRIDGE; a fine Engraving. The work of the RENNIES.

In royal 8vo, with very elaborate Plates (folded), £1 10s.

BRIDGES, SUSPENSION. — An Account, with Illustrations, of the Suspension Bridge across the River Danube, by Wm. T. CLARK, F.R.S.

In 4 vols., royal 8vo, bound in 3 vols., half-morocco, price £4 10s.

BRIDGES. — THE THEORY, PRACTICE, AND ARCHITECTURE OF BRIDGES OF STONE, IRON, TIMBER, AND WIRE; with Examples on the Principle of Suspension; Illustrated by 138 Engravings and 92 Woodcuts.

In one large 8vo volume, with explanatory Text, and 68 Plates comprising details and measured dimensions. Bound in half-morocco, uniform with the preceding work, price £2 10s.

BRIDGES. — SUPPLEMENT TO "THE THEORY, PRACTICE, AND ARCHITECTURE OF BRIDGES OF STONE, IRON, TIMBER, WIRE, AND SUSPENSION."

1 large folio Engraving, price 7s. 6d.

BRIDGE across the Thames.—SOUTHWARK IRON BRIDGE.

1 large folio Engraving, price 5s.

BRIDGE across the Thames. — WATERLOO STONE BRIDGE.

1 very large Engraving, price 5s.

BRIDGE across the Thames. — VAUXHALL IRON BRIDGE.

1 very large Engraving, price 4s. 6d.

BRIDGE across the Thames.—HAMMERSMITH SUSPENSION BRIDGE.

1 large Engraving, price 4s. 6d.

BRIDGE (the UPPER SCHUYLKILL) at PHILADELPHIA, the greatest known span of one arch, covered.

1 large Engraving, price 3s. 6d.

BRIDGE (the SCHUYLKILL) at PHILA-DELPHIA, covered.

1 large Engraving, price 3s. 6d.

BRIDGE. — ON THE PRINCIPLE OF SUS-PENSION, by Sir I. BRUNEL, in the ISLAND OF BOURBON.

1 large Engraving, price 4s.

BRIDGE. — PLAN and ELEVATION of the PATENT IRON BAR BRIDGE over the River Tweed, near Berwick.

84 Plates, folio, £1 1s., boards.

BRIGDEN, R. — Interior Decorations, Details, and Views of Sefton Church, Lancashire, erected in the reign of Henry VIII.

John Weale, 59, High Holborn, London, W.C.

MR. WEALE'S WORKS ON ARCHITECTURE, ENGINEERING, FINE ARTS, &c.

1 large Engraving, price 3s. 6d.

BRITTON'S (John) VIEWS of the WEST FRONTS of 14 ENGLISH CATHEDRALS.

1 large Engraving in outline, price 2s. 6d.

BRITTON'S (John) PERSPECTIVE VIEWS of the INTERIOR of 14 CATHEDRALS.

In 4to, 2s. 6d.

BRODIE, R., C.E. — Rules for Ranging Rail- way Curves, with the Theodolite, and without Tables.

1 large Engraving, price 4s. 6d.

BROWN'S (Capt. S.) CHAIN PIER at Brighton, with Details.

The Text in one large volume 8vo, and the Plates, upwards of 70 in number, in an atlas folio volume, very neatly half-bound, £2 10s.

BUCHANAN, R. — PRACTICAL ESSAYS ON MILL WORK AND OTHER MACHINERY; with Examples of Tools of modern invention; first published by ROBERT BUCHANAN, M.E.; afterwards improved and edited by THOMAS TREDGOLD, C.E.; and re-edited, with the improvements of the present age, by GEORGE RENNIE, F.R.S., C.E., &c., &c. The whole forming 70 Plates, and 103 Woodcuts. John Weale, 59, High Holborn, London, W.C.

Text in royal 8vo, and Plates in imperial folio, 18s.

BUCHANAN, R. — SUPPLEMENT. — PRACTICAL EXAMPLES ON MODERN TOOLS AND MACHINES; a Supplementary Volume to Mr. RENNIE'S edition of BUCHANAN "On Mill-Work and Other Machinery," by TREDGOLD. The work consists of 18 Plates.

In 8vo, with Plates, 2nd Edition, 1s. 6d.

BURN, C., C.E.—On Tram and Horse Railways.

In one volume, 4to, 21 Plates, half-bound in morocco, £1 1s.

BURY, T., Architect. — Examples of Ancient Ecclesiastical Woodwork.

7s. 6d.

CALCULATOR (THE): Or, TIMBER MER- CHANT'S AND BUILDER'S GUIDE. By WILLIAM RICHARDSON and CHARLES GANE, of Wisbeach.

In 8vo, Plates, cloth boards, 7s. 6d.

CALVER, E. K., R.N.—THE CONSERVATION AND IMPROVEMENT OF TIDAL RIVERS.

In 8vo, Woodcuts, 1s 6d.

CALVER, E.K., R.N.—ON THE CONSTRUC- TION AND PRINCIPLE OF A WAVE SCREEN, designed for the Formation of Harbours of Refuge.

In 4to, half-bound, price £1 5s.

CARTER, OWEN B., Architect.—A SERIES OF THE ANCIENT PAINTED GLASS OF WINCHESTER CATHEDRAL, Examples of. 28 Coloured Illustrations.

In 4to, 17 Plates, half-bound, 7s. 6d.

CARTER, OWEN B., Architect. —ACCOUNT OF THE CHURCH OF ST. JOHN THE BAPTIST, at Bishopstone, with Illustrations of its Architecture.
John Weale, 59, High Holborn, London, W.C.

MR. WEALE'S WORKS ON ARCHITECTURE, ENGINEERING, FINE ARTS, &c.

In 4to, with 19 Engravings, £1 1s.

CHATEAUNEUF, A. de, Architect.—Architectura Domestica; a Series of very neat examples of Interiors and Exteriors of residences in the Italian style.

Large 4to, in half-red morocco, price £1 8s.

CHIPPENDALE, INIGO JONES, JOHNSON, LOCK, and PETHER.—Old English and French Ornaments; comprising 244 designs on 105 Plates of elaborate examples of Hall Glasses, Picture Frames, Chimney-pieces, Ceilings, Stands for China, Clock and Watch Cases, Girandoles, Brackets, Grates, Lanterns, Ornamental Furniture, Ornaments for brass workers and silver workers, real ornamental iron work Patterns, and for carvers, modellers, &c., &c., &c.

4to, third Edition with additions, price £1 11s. 6d.

CLEGG, SAM., C.E.—A PRACTICAL TREATISE ON THE MANUFACTURE AND DISTRIBUTION OF COAL GAS, Illustrated by Engravings from Working Drawings, with General Estimates.

In 4to, Plates, and 76 Woodcuts, boards, price 6s.

CLEGG, SAM., C.E.—ARCHITECTURE OF MACHINERY. An Essay on Propriety of Form and Proportion. For the use of Students and Schoolmasters.

In 8vo, 1s.

COLBURNS, Z.—On Steam Boiler Explosions.

One very large Engraving, price 4s. 6d.

CONEY'S (J.) Interior View of the Cathedral Church of St. Paul.

In 4to, on card board, 1s.

COWPER, C.—Diagram of the Expansion of Steam.

In one vol. 4to, with 20 Folding Plates, price £1 1s.

CROTON AQUEDUCT. — Description of the New York Croton Aqueduct, in 20 large detailed and engineering explanatory Plates, with text in the English, German, and French languages, by T. SCHRAMKE, C.E.

In demy 12mo, cloth, extra bound and lettered, price 4s.

DENISON.—A Rudimentary Treatise on Clocks and Watches, and Bells; with a full account of the Westminster Clock and Bells, by EDMUND BECKET DENISON, M.A., Q.C. Fourth Edition re-written and enlarged, with Engravings.

In royal 4to, cloth boards, price £1 11s. 6d.

DOWNES, CHARLES, Architect.—Great Exhibition Building. The Building erected in Hyde Park for the Great Exhibition, 1851; 28 large folding Plates, embracing Plans, Elevations, Sections, and Details, laid down to a large scale, and the Working and Measured Drawings.

DRAWING BOOKS.—Showing to Students the superior method of Drawing and Shadowing.

DRAWING BOOK.—COURS ELEMENTAIRES DE LAVIS APPLIQUÉ À L'ARCHITECTURE; folio volume, containing 40 elaborately engraved Plates, in shadows and tints, very finely executed, by the best artists in France. £2. Paris.

John Weale, 59, High Holborn, London, W.C.

MR. WEALE'S WORKS ON ARCHITECTURE, ENGINEERING, FINE ARTS, &c.

DRAWING BOOK. — COURS ÉLÉMEN-
TAIRES DE LAVIS APPLIQUÉ À MÉCHANIQUE)
folio volume, containing 50 elaborately engraved Plates, in shadows
and tints, very finely executed, by the best artists in France.
£2 10s. Paris.

DRAWING BOOK. — COURS ÉLÉMEN-
TAIRES DE LAVIS APPLIQUÉ À ORNEMENTA-
TION; folio volume, containing 20 elaborately engraved Plates, in
shadows and tints, very finely executed, by the best artists in
France. £1. Paris.

DRAWING BOOK. — ÉTUDES PROGRES-
SIVES ET COMPLÈTES D'ARCHITECTURE DE
LAVIS, par J. B. TRITON; large folio, 24 fine Plates, comprising
the Orders of Architecture, mouldings, with profiles, ornaments,
and forms of their proportion, art of shadowing doors, balusters,
parterres, &c., &c., &c. £1 4s. Paris.

In 12mo, cloth boards, lettered, price 5s.

ECKSTEIN, G. F. — A Practical Treatise on
Chimneys; with remarks on Stoves, the consumption of
Smoke and Coal, Ventilation, &c.

Plates, imperial 8vo, price 7s.

ELLET, CHARLES, C. E., of the U. S. — Report
on the Improvement of Kanawha, and incidentally of the
Ohio River, by means of Artificial Lakes.

In 8vo, with Plates, price 12s.

EXAMPLES of Cheap Railway Making,
American and Belgian.

In one vol. 4to, 49 Plates, with dimensions, extra cloth boards,
price 21s.

EXAMPLES for Builders, Carpenters, and
Joiners; being well-selected Illustrations of recent Modern
Art and Construction.

With Engravings and Woodcuts, price 12s.

FROME, Lieutenant-Colonel, R.E. — Outline of
the Method of conducting a Trigonometrical Survey, for the
Formation of Topographical Plans; and Instructions for filling in
the Interior Detail, both by Measurement and Sketching; Military
Reconnaissances, Levelling, &c., &c., together with Colonial Surveying.

In 4to, with Plates, price 7s. 6d.

FAIRBAIRN, W., C.E., F.R.S. — ON
WATER-WHEELS, WITH VENTILATED BUCKETS.

In royal 8vo, with Plates and Woodcuts, Second Edition, much
improved, price, in extra cloth boards, 16s.

FAIRBAIRN, W., C.E., F.R.S. — ON THE
APPLICATION OF CAST AND WROUGHT IRON TO
BUILDING PURPOSES.

In imperial 8vo, with fine Plates, a re-issue, price 16s., or 21s. in
half-morocco, gilt edges,

FERGUSSON'S (J.) Essay on the Ancient Topo-
graphy of Jerusalem, with restored Plans of the Temple, &c.

In 8vo, sewed in wrapper, price 2s.

GILL, J. — ESSAY ON THE THERMO DY-
NAMICS OF ELASTIC FLUIDS, by JOSEPH GILL,
with Diagrams.

John Weale, 59, High Holborn, London, W.C.

MR WEALE'S WORKS ON ARCHITECTURE, ENGINEERING, FINE ARTS, &c.

Plates, 8vo, boards, 5s.

GWILT, JOSEPH, Architect.—TREATISE ON THE EQUILIBRIUM OF ARCHES.

In 8vo, cloth boards, with 8 Plates, 4s. 6d.

HAKEWELL, S. J.—Elizabethan Architecture ; illustrated by parallels of Dorton House, Hatfield, Longleat, and Wollaton, in England, and the Palazzo Della Cancellaria at Rome.

8vo, with a Map. 1s.

HAMILTON, P. S., Barrister-at-Law, Halifax Nova Scotia—Nova Scotia considered as a Field for Emigration.

In Imperial 8vo, Third Edition, with additions, 11 Plates, cloth boards, 8s.

HART, J., On Oblique Bridges. — A Practical Treatise on the Construction of Oblique Arches.

In 4to, with Woodcuts, 3s. 6d.

HEALD, GEORGE, C.E.—System of Setting Out Railway Curves.

Royal 8vo, Plates and Woodcuts, price 12s. 6d.

HEDLEY, JOHN. — Practical Treatise on the Working and Ventilation of Coal Mines, with Suggestions for Improvements in Mining.

Two Vols., demy 12mo, in cloth extra boards and lettered, price 12s. 6d.

HOMER. — The Iliad and Odyssey, with the Hymns of Homer, Edition with an accession of English notes by the Rev. T. H. L. LEARY, M.A.

In 8vo, with Engravings, cloth boards, Third Edition, 10s. 6d.

HOPKINSON, JOSEPH, C.E.—The Working of the Steam Engine Explained by the use of the Indicator.

In 8vo, in boards, 18s.

HUNTINGTON, J. B., C.E. — TABLES and RULES for Facilitating the Calculation of Earthwork, Land, Curves, Distances, and Gradients, required in the Formation of Railways, Roads, and Canals.

Separate from the above, price 3s.

HUNTINGTON, J. B., C.E. — THE TABLES OF GRADIENTS.

10 Plates, 8vo, bound, 5s.

INIGO JONES.—Designs for Chimney Glasses and Chimney Pieces of the Time of Charles the 1st.

In a sheet, 2s.

IRISH.—Plantation and British Statute Measure (comparative Table of), so that English Measure can be transferred into Irish, and vice versa.

In 4to, with 8 Engravings, in a wrapper, 6s.

IRON. — ACCOUNT OF THE CONSTRUCTION OF THE IRON ROOF OF THE NEW HOUSES OF PARLIAMENT, with elaborate Engravings of details.

In imperial 4to, with 50 Engravings, and 2 fine Woodcuts, halfbound in morocco, £1 4s.

IRON. — DESIGNS OF ORNAMENTAL GATES, LODGES, PALISADING, AND IRON-WORK OF THE ROYAL PARKS, with some other Designs.

John Weale, 59, High Holborn, London, W.C.

MR. WEALE'S WORKS ON ARCHITECTURE, ENGINEERING, FINE ARTS, &c.

In 4to, with 10 Plates, 12s.

JEBB'S, Colonel, Modern Prisons.—Their Construction and Ventilation.

In 3 vols. 8vo, with 26 elaborate Plates, cloth boards, £2 2s.

JONES, Major-Gen. Sir John, Bart. — Journal of the Sieges carried on by the Army under the Duke of Wellington in Spain, between the years 1811 and 1814, with an Account of the Lines of Torres Vedras. By Major-Gen. Sir JOHN T. JONES, Bart., K.C.B. Third Edition, enlarged and edited by Lieut.-General Sir HARRY D. JONES, Bart.

16mo, cloth boards, 2s. 6d.

KENNEDY AND HACKWOOD'S Tables for Setting out Curves.

In 4to, 37 Plates, half-cloth boards, 9s.

KING, THOMAS.—The Upholsterer's Guide; Rules for Cutting and Forming Draperies, Valances, &c.

Illustrated by large Draughts and Engravings. In 1 volume 4to, text, and a large atlas folio volume of Plates, half-bound, £6 6s.

KNOWLES, JOHN, F.R.S.—The Elements and Practice of Naval Architecture; or, A Treatise on Ship Building, theoretical and practical, on the best principles established in Great Britain; with copious Tables of Dimensions, Scantlings, &c. The Third Edition, with an Appendix, containing the principles of constructing the Royal and Mercantile Navies, by Sir ROBERT SEPPINGS.

41 Plates of a fine and an elaborate description in large atlas folio half-bound, £2 12s. 6d.; with the text half-bound in 4to.

LOCOMOTIVE ENGINES. — The Principles and Practice and Explanation of the Machinery of Locomotive Engines in operation.

In 12mo, sewed, 1s.

MAIN, Rev. ROBERT. — An Account of the Observatories in and about London.

4to, in boards, 15s.

MANUFACTURES AND MACHINERY. — Progress of, in Great Britain, as exhibited chiefly in Chronological notices of some Letters Patent granted for Inventions and Improvements, from the earliest times to the reign of Queen Anne.

16mo, 2s. 6d.

MAY, R. C., C.E.—Method of setting out Railway Curves.

Imperial 4to, with fine Illustrations, extra cloth boards, £1 5s., or half-bound in morocco, £1 11s. 6d.

METHVEN, CAPTAIN ROBERT.—THE LOG OF A MERCHANT OFFICER, Viewed with Reference to the Education of Young Officers and the Youth of the Merchant Service. By ROBERT METHVEN, Commander in the Peninsular and Oriental Company's Service.

In royal 8vo, 1s. 6d.

METHVEN, CAPTAIN ROBERT.—NARRATIVES WRITTEN BY SEA COMMANDERS, ILLUSTRATIVE OF THE LAW OF STORMS. The "Blenheim" Hurricane of 1851, with Diagrams.

Part 1, large 8vo, 5s. Part 2, in preparation.

MURRAY, JOHN, C.E. — A Treatise on the Stability of Retaining Walls, elucidated by Engravings and Diagrams.

John Weale, 59, High Holborn, London, W.C.

MR. WEALE'S WORKS ON ARCHITECTURE, ENGINEERING, FINE ARTS, &c.

On a large folio sheet, price 2s. 6d.

NEVILLE, JOHN, C.E., M.R.I.A. — OFFICE HYDRAULIC TABLES: for the use of Engineers engaged in Water Works, giving the Discharge and Dimensions of River Channels and Pipes.

In 8vo, Second and much Improved Edition, with an Appendix, cloth boards, price 16s.

NEVILLE, JOHN, C.E., M.R.I.A. — HYDRAULIC TABLES, COEFFICIENTS, AND FORMULÆ; for Finding the Discharge of Water from Orifices, Notches, Weirs, Pipes, and Rivers, with Extensive Additions, New Formulæ, Tables, and General Information on Rain-Fall Catchment-Basins, Drainage, Sewerage, Water Supply for Towns and Mill Power.

On 33 folio Plates, 12s.

ORNAMENTS. — Ornaments displayed on a full size for Working, proper for all Carvers, Painters, &c., containing a variety of accurate examples of foliage and friezes.

Plates, 8vo, 2s. 6d.

O'BRIEN'S, W., C.E. — Prize Essay on Canals and Canal Conveyance.

In demy 8vo, cloth, boards, 12s.

PAMBOUR, COUNT DE. — STEAM ENGINE; the Theory of the Proportions of Steam Engines, and a series of practical formulæ.

In 8vo, cloth, boards, with Plates. a second edition, 18s.

A PRACTICAL TREATISE ON LOCOMOTIVE ENGINES UPON RAILWAYS. — With practical Tables and an Appendix, showing the expense of conveying Goods by means of Locomotives on Railroads. By COUNT F. M. G. DE PAMBOUR.

4to, 72 finely executed Plates, in cloth, £1 16s.

PARKER, CHARLES, Architect, F.I.B.A. — The Rural and Villa Architecture of Italy, portraying the several very interesting examples in that country, with Estimates and Specifications for the application of the same designs in England; selected from buildings and scenes in the vicinity of Rome and Florence, and arranged for Rural and Domestic Buildings generally.

Price, complete, £2 2s. In 4to.

POLE, WILLIAM, M. Inst., C. E. — CORNISH PUMPING ENGINE; designed and constructed at the Hayle Copper House in Cornwall, under the superintendence of CAPTAIN JENKINS; erected and now on duty at the Coal Mines of Languin, Department of the Loire Inférieur, Nantes. Nine elaborate Drawings, historically and scientifically described.

With Plate. 10s. 6d.

AN ANALYTICAL INVESTIGATION OF THE ACTION OF THE CORNISH PUMPING ENGINE. — This Third Part sold separately from above.

2s. bound in 4to size.

PORTFOLIO OF ENGINEERING ENGRAVINGS. — Useful to Students as a Text Book, or a Drawing Book of Engineering and Mechanics; being a series of Practical Examples in Civil, Hydraulic, and Mechanical Engineering. Fifty Engravings to a scale for drawing.

John Weale, 59, High Holborn, London, W.C.

MR. WEALE'S WORKS ON ARCHITEC-
TURE, ENGINEERING, FINE ARTS, &c.

In royal 8vo, uniform with the preceding, 9s., with Charts and Woodcuts. The work together in 2 vols., £1 1s.

REID, Major-General Sir W., F.R.S., &c. — THE PROGRESS OF THE DEVELOPMENT OF THE LAW OF STORMS AND OF THE VARIABLE WINDS, with the practicable application of the subject to navigation.

Illustrated with 17 Plates, Third Edition, 8vo, cloth, 7s. 6d.

RICHARDSON, C. J., Architect. — A Popular Treatise on the Warming and Ventilation of Buildings: showing the advantage of the improved system of Heated Water Circulation. And a method to effect the combination of large and small pipes to the same apparatus, and ventilating buildings.

Bound in 2 vols., very neat, half-morocco, gilt tops, price £1S.

RENNIE'S, Sir JOHN, F.R.S., Work on the Theory, Formation, and Construction of British and Foreign Harbours, Docks, and Naval Arsenals. This great work may now be had complete, 20 parts and supplement, price £16.

In 8vo, 2s.

RÉVY, J. L., C.E. — THE PROGRESSIVE SCREW AS A PROPELLER IN NAVIGATION.

12mo, cloth boards. 3s. 6d.

SIMMS, F. W. — Treatise on the principal Mathematical and Drawing Instruments employed by the Engineer, Architect, and Surveyor; with a description of the Theodolite, together with Instructions in Field Works.

4to, with fine Plates, a New Edition, extended, sewed, 5s.

SMITH, C. H., Sculptor.—Report and Investiga- tion into the Qualifications and Fitness of Stone for Building Purposes.

In 1 vol. 8vo, in boards, 7s. 6d.

SMITH'S, Colonel of the Madras Engineers, Observations on the Duties and Responsibilities Involved in the Management of Mines.

8vo, cloth boards. with Index Map, 5s.

SOPWITH, THOMAS, F.R.S. — THE AWARD OF THE DEAN FOREST COMMISSIONERS AS TO THE COAL AND IRON MINES.

16 large folio Plates, £1 4s. Separately, 2s. each.

SOPWITH, THOMAS, F.R.S.—SERIES OF ENGRAVED PLANS OF THE COAL AND IRON MINES.

12 Plates, 4to, 6s. in a wrapper.

STAIRCASES, HANDRAILS, BALUSTRADES, AND NEWELS OF THE ELIZABETHAN AGE, &c.— Consisting of — 1. Staircase at Audley-end Old Manor House, Wilts; 2. Charlton House, Kent; 3. Great Ellingham Hall; Norfolk; 4. Dorfold; Cheshire; 5. Charterhouse; 6. Oak Staircase at Clare Hall, Cambridge; 7. Cromwell Hall, Highgate; 8. Ditto; 9. Catherine Hall, Cambridge; 10. Staircase by Inigo Jones at a house in Chandos Street; 11. Ditto at East Sutton; 12. Ditto, ditto. Useful to those constructing edifices in the early English domestic style.

Large atlas folio Plates, price £2 2s.

STALKARTT, M., N.A. — Naval Architecture; or, The Rudiments and Rules of Ship Building: exemplified in a Series of Draughts and Plans. No text.

John Weale, 59, High Holborn, London, W.C.

MR. WEALE'S WORKS ON ARCHITEC-
TURE, ENGINEERING, FINE ARTS, &c.

With Illustrative Diagrams. In 8vo, 7s. 6d.

STEVENSON'S, THOMAS, C.E., of Edinburgh,
Description of the Different kinds of Lighthouse Apparatus.

8vo, 2s. 6d.

STEVENSON, DAVID, C.E., of Edinburgh. —
Supplement to his Work on Tidal Rivers.

Text in 4to, and large folio Atlas of 75 Plates, half-cloth boards,
£2 12s. 6d.

STEAM NAVIGATION. — Vessels of Iron and
Wood; the Steam Engine; and on Screw Propulsion. By
WM. FAIRBAIRN, F.R.S., of Manchester; Messrs. FORRESTER,
M.I.C.E., of Liverpool; JOHN LAIRD, M.I.C.E., of Birkenhead;
OLIVER LANG, (late) of Woolwich; Messrs. SEAWARD, Lime-
house, &c. &c. &c. Together with Results of Experiments on the
Disturbance of the Compass in Iron-built Ships. By G. B. AIRY,
M.A., Astronomer Royal.

10s.

ST. PAUL'S CATHEDRAL, LONDON, SEC-
TION OF. — The Original Splendid Engraving by J.
GWYN, J. WALE, decorated agreeably to the original intention
of Sir Christopher Wren; a very fine large print, showing distinctly
the construction of that magnificent edifice.

Size of Plate 4½ feet in height, 10s.

ST. PAUL'S CATHEDRAL, LONDON, GREAT
PLAN. — J. WALE and J. GWYN'S GREAT PLAN,
accurately measured from the Building, with all the Dimensions
figured and in detail, description of Compartments by engraved
Writing.

Second Edition, greatly enlarged, royal 8vo, with Plates, cloth
boards, price 16s.

STRENGTH OF MATERIALS.—FAIRBAIRN,
WILLIAM, C.E., F.R.S., and of the Legion of Honour of
France. On the application of Cast and Wrought Iron to Building
Purposes.

With Plates and Diagrams. New Edition. The work complete
in 2 vols., bound in 1 vol., price, in cloth boards, 16s. The
second portion of the work, containing Mr. Hodgkinson's Experi-
mental Researches, may be had separately, price 9s.

STRENGTH OF MATERIALS.—HODGKIN-
SON, EATON, F.R.S., AND THOMAS TREDGOLD,
C.E. A PRACTICAL ESSAY ON THE STRENGTH OF CAST
IRON AND OTHER METALS; intended for the assistance of
engineers, ironmasters, millwrights, architects, founders, smiths
and others engaged in the construction of machines, buildings, &c'
By EATON HODGKINSON, F.R.S.

To be published in 1861, in crown 8vo, bound for use.

STRENGTH OF MATERIALS.—POLE, WIL-
LIAM, C.E., F.R.S.—Tables and popular explanations of
the Strength of Materials, of Wrought and Cast Iron with other
metals, for structural purposes; developing in a systematic form,
the strengths, bearings, weights, and forms of these materials, whether used as girders or arches, for the construction of bridges and
viaducts, public buildings, domestic mansions, private buildings,
columns or pillars, breasummmiers for warehouses, shops, working
and manufacturing factories, &c. &c. &c. The whole rendered of
easy reference for architects, builders, civil and mechanical engineers, millwrights, ironfounders, &c. &c. &c., and forming Ready
Reckoner or Calculater.

John Weale, 59, High Holborn, London, W.C.

MR. WEALE'S WORKS ON ARCHITECTURE, ENGINEERING, FINE ARTS, &c.

90 very elaborately drawn Engravings. In large 4to, neatly half-bound and lettered, £1 1s. A few copies on large imperial size, extra half-binding. £1 11s. 6d.

TEMPLE CHURCH.—The Architectural History
and Architectural Ornaments, Embellishments, and Painted Glass, of the Temple Church, London.

Part I., with 26 Engravings on Wood and Copper, in cloth boards, 4to, 15s.

THAMES TUNNEL.—A Memoir of the several
Operations and the Construction of the Thames Tunnel, from Papers by the late Sir ISAMBARD BRUNEL, F.R.S., Civil Engineer.

Fourth Edition, with a Supplementary Addition, large 8vo, 12s. 6d.

THOMAS (LYNALL), F.R.S.L.—Rifled Ordnance.
—A Practical Treatise on the Application of the Principle of the Rifle to Guns and Mortars of every calibre; to which is added a New Theory of the Initial Action and Force of Fixed Gunpowder plates.

In 4to, complete, cloth, Vol. I., with Engravings, £1 10s.; Vol. II., ditto, £1 8s.; Vol. III., ditto, £2 12s. 6d.

TRANSACTIONS OF THE INSTITUTION
OF CIVIL ENGINEERS.

8 vols., numerous Engravings of Sections of Coal Mines, &c., large folding Plates, several of which are coloured, in large 8vo, half-bound in calf, price £1 1s. per volume.

TRANSACTIONS OF THE NORTH OF
ENGLAND INSTITUTE OF MINING ENGINEERS.—Commencing in 1852, and continued to 1860.

A New Edition revised by the translator, and with additional Plates, in demy 12mo, India proof Plates and Vignettes, half-bound in morocco, gilt tops, price 12s. Only 25 printed on India paper.

VITRUVIUS. — The Architecture of Marcus
Vitruvius Pollio in 10 Books. Translated from the Latin by JOSEPH GWILT, F.S.A., F.R.A.S.

In 4to, with Plates, 7s. 6d.

WALKER'S, THOMAS, Architect. — Account
of the Church at Stoke Golding.

£1 10s.

WEALE'S QUARTERLY PAPERS ON EN-
GINEERING. — Vol. VI. (Parts 11 and 12 completing the work.) Comprising, "On the Principles of Water Power." Plates. Experiments on Locomotive Engines. Coloured Plates. On Naval Arsenals. On the Mode of Forming Foundations under water and on bad ground. Plates. On the Improvement of the River Medway and of the Fort and Arsenal of Chatham. On the Improvement of Portsmouth Harbour. An Analysis of the Cornish Pumping. Plates. On Water Wheels. Plates.

Text in 8vo. cloth boards, and Plates in atlas folio, in cloth, 16s.

WHITE'S, THOMAS, N.A., Theory and Prac-
tice of Ship Building.

In 8vo. with a large Sectional Plate, 1s. 6d.

WHICHCORD, JOHN, Architect. —
OBSERVATIONS ON KENTISH RAG STONE AS A BUILDING MATERIAL.
John Weale, 59, High Holborn, London, W.C.

MR. WEALE'S WORKS ON ARCHITEC-
TURE, ENGINEERING FINE ARTS, &c.

4to, coloured Plates, in half-morocco, 7s 6d.

WHICHCORD, JOHN, Architect.—HIS-
TORY AND ANTIQUITIES OF THE COLLEGIATE
CHURCH OF ALL SAINTS, MAIDSTONE.

In 4to, 6s.

WICKSTEED, THOMAS, C.E. — AN EXPE-
RIMENTAL INQUIRY CONCERNING THE RELA-
TIVE POWER OF, AND USEFUL EFFECT PRODUCED
BY, THE CORNISH AND BOULTON & WATT PUMPING
ENGINES, and Cylindrical and Waggon-Head Boilers.

In 8vo, 1s.

WICKSTEED, THOMAS, C.E. — FURTHER
ELUCIDATION OF THE USEFUL EFFECTS OF
CORNISH PUMPING ENGINES; showing the average work-
ing for long periods, &c., &c., &c.

£2 2s.

WICKSTEED, THOMAS, C.E. — THE
ELABORATELY ENGRAVED ILLUSTRATIONS OF
THE CORNISH AND BOULTON & WATT ENGINES erected
at the East London Water Works, Old Ford. Eight large atlas
folio very fine line engravings by GLADWIN, from elaborate
drawings made expressly by Mr. WICKSTEED; folio, together
with a 4to explanation of the plates, containing an engraving, by
LOWRY, of Harvey and West's patent pump-valve, with speci-
fication.

With numerous Woodcuts.

WILLIAMS, C. WYE, Esq., M. Inst. C. E.—
THE COMBUSTION OF COAL AND THE PREVEN-
TION OF SMOKE, chemically and practically considered.

Imperial 8vo, with a Portrait, 2s. 6d.

WILLIAMS, C. WYE, Esq, M. Inst. C. E. —
PRIZE ESSAY ON THE PREVENTION OF THE
SMOKE NUISANCE, with a fine portrait of the Author.

With 3 Plates, containing 51 figures, 4to, 5s.

WILLIS, REV. PROFESSOR, M.A.—A
system of Apparatus for the use of Lecturers and Experi-
menters in Mechanical Philosophy.

In 4to, bound, with 26 large plates and 17 woodcuts, 12s.

WILME'S MANUALS. — A MANUAL OF
WRITING AND PRINTING CHARACTERS, both
ancient and modern.

Maps and Plans, in 4to, plates coloured, half-bound morocco, £2.

WILME'S MANUALS. — A HANDBOOK
FOR MAPPING, ENGINEERING, AND ARCHITEC-
TURAL DRAWING.

Three Vols., large 8vo, £3.

WOOLWICH. — COURSE OF MATHEMA-
TICS. This course is essential to all Students destined
for the Royal Military Academy at Woolwich.

8vo, 1s.

YULE, MAJOR-GENERAL.—ON BREAK-
WATERS AND BUOYS of VERTICAL FLOATS.
John Weale, 59, High Holborn, London, W.C.

FOREIGN WORKS, KEPT IN STOCK AS FOLLOWS:—

Large folio, 32 plates, some coloured, and 12 woodcuts, 50 francs. £2 10s.

ARCHITECTURE SUISSE.—Ou Choix de Maisons Rustiques des Alpes du Canton de Berne, par GRAF-FINRIED et STÜRLER, Architectes. Berne, 1841.

Small folio, 52 most interesting and explanatory plates of Public Works, Bridges, Iron Works, &c., &c., &c., very neatly half-bound in morocco, £1 10s.

BAUERNFEIND, CARL MAX.—VORLEGE-BLAETTER ZUR BRUCKENBAU KUNDE. München.

Large folio, 36 plates of Byzantine capit'ls, 12s.

BYZANTINISCHE CAPITAELER.—München.

Second edition, 126 plates, large folio, best Paris edition, 100 f., printed on fine paper, half-cloth boards, £4 4s.

CALLIAT, VICTOR, ARCT.—Parallèle des Maisons de Paris, construites depuis 1830 jusqu'à nos jours.—1857.

Large folio, 60 francs, 60 plates, and several vignettes, £2 8s.

CANÉTO, F.—Sainte-Marié d'Auch. Atlas Monographiquie de Cette Cathédrale. The Plates consist principally of outline drawings of the Painted Glass Windows in this Cathedral.

120 plates, elegant in half-morocco extra, interleaved, £5 15s. 6d.

CASTERMAN, A.—PARALÈLLE des MAISONS de BRUXELLES et des PRINCIPALES VILLES de la BELGIQUE, construites depuis 1830 jusqu'à nos jours, représentés en plans, élévations, coupes et détails intérieurs et extérieurs.—Paris.

Small folio, 48 plates of edifi es £1 1s.

DEGEN, L. — LES CONSTRUCTIONS EN BRIQUES, composées et publiées. 8 livraisons.—1859.

Small folio, 48 plates of houses, parts of houses, details of all kinds of singularly beautiful woodwork, coloured plates in imitation of the objects given, £1 1s.

DEGEN, L.—LES CONSTRUCTIONS ORNAMENTALES EN BOIS, 8 livraisons.

In 3 very large folio parts, 85 fine plates, £1 11s. 6d.

GAERTNER, F. V.—The splendid works of M. GAERTNER of Munich, drawn to a very large size, consisting of the library in plans, elevations, interiors, details, and sections, and coloured ornaments. The church, with details, ornaments, &c.—München.

Small folio, 86 fine plates of the Architecture, ornament, and detail of the houses and churches of Germany during the middle age, very neatly half-bound in morocco, £2 12s. 6d.

KALLENBACH, C. C.—Chronologie der Deutsch-Mittelalterlichen Baukunst.—München. Fine Work.

The works of the great master KLENZIE of Munich, in 5 parts very large folio, 50 plates of elevations, plans, sections, details and ornaments of his public and private buildings executed in Munich and St. Petersburg, £2 2s.

KLENZE, LEO VON. — Sammlung Arthitectonisher Entwürfe, fur die Ausführung bestimmt oder wirklich ausgeführt. Published in Munich.

John Weale, 59, High Holborn, London, W.C.

FOREIGN WORKS KEPT IN STOCK AS
FOLLOWS:—

Upwards of 100 plates, large 4to, £2 12s. 6d.

PETIT, VICTOR.—CHATEAUX DE FRANCE.
Architecture Pittoresque, ou Monuments des quinzième et seizième siècles. Paris.

Livraisons 1 à 13, very finely executed plates, large imperial folio, £5 8s.

CHATEAUX DE LA VALLÉE DE LA LOIRE DES XV, XVI, ET COMMENCEMENT DU XVII SIECLE.—Paris, 1857—60.

4to, 96 plates, 72f.; £2 10s.

RECUEIL DE SCULPTURES GOTHIQUE.—
Dessinées et gravées à l'eau forte d'après les plus beaux monuments construits en France depuis le onzème jusqu'au quinzième siècle, par ADAMS, Inspecteur des travaux de la Sainte Chapelle. Paris, 1856.

4 parts are published, price 14s.

RAMÉE.—HISTOIRE GÉNÉRALE DE L'ARCHITECTURE. L'histoire générale de l'Architecture, par DANIEL RAMÉE, forme 2 vol. grande in 8vo, publiés en 8 fascicules.

5 vols., large 8vo, numerous fine woodcuts, half morocco.

VIOLET-LE-DUC. — DICTIONNAIRE RAISONNE, de l'Architecture Française du quinzième au seizième siècle. Paris, 1854-8.

2 vols., extra imperial folio, price £6 16s. 6d.

BADIA D'ALTACOMBA.—Storia e Descrizione della Antico Sepolchro del Reali di Savoia, fondita da Amedio III. rinnovata da Carlo Felice e Maria Christina.

79 livraisons in large 4to, 200 engravings, £8 18s. 6d.

BELLE ARTI.—Il Palazzo Ducale di Venezia, illustrato da Francesco Zanotto. Venezia, 1816—1858.

2 vols. large 4to, 62 very neatly engraved outline Plates, £1 5s.

CANOVA.—Le Tombe ed i Monumenti Illustri d'Italia. Milano.

2 vols. 4to, 67 elaborate Plates, £1 16s.

CAVALIERI SAN-BERTOLO (NICOLA).— ISTITUZIONI DI ARCHITETTURA STATICA E IDRAULICA. Mantova.

2 vols. imperial folio, in parts of eight divisions, &c., New and much Improved Edition, comprising 259 Plates of the Public Buildings of Venice, plans, elevations, sections, and details, £8 18s. 6d.

CICOGNARA (COUNT).—Le Fabbriche ei Monumenti Copiicoi di Venezia, illustrati da L. Cicognara, da A. Diedo, e da G. A. Selva, edizione con copiose note ed aggiunte di Francesco Zanotto, arricchita di nuove tavole e della Versione Francese. Venezia nello stab. naz. di G. Antonelli a spese degli edit. G. Antonelli e Luciano Basadonna, 1858. The elaborately descriptive text is in French and Italian, beautifully printed.
Copies elegantly half-bound in morocco, extra gilt, library copy and interleaved, £12 12s. Venezia, 1858.

Full, Portrait, and 147 Plates, consisting of subjects of public buildings, executed at Verona, plans, elevations, sections, details, and ornaments, with some executed works at Venice, &c., £4 4s.

FABBRICHE.—CIVILI ECCLESIASTICHE E MILITARI DI MICHELE SAN MICHELE disegnate ed incise da RONZANI FRANCESCO e L. GIROLAMO.

John Weale, 59, High Holborn, London, W.C.

FOREIGN WORKS KEPT IN STOCK AS FOLLOWS.

Large folio, containing a profusion of Plates of the palaces, theatres, hôtel de villes, and other public buildings in several parts of Italy. Elegantly half-bound in red morocco, extra gilt and interleaved, £6 6s.

FABBRICHE.—E DISEGNI D'ANTONIO
DIEDO, NOBILE VENETO. Venezia.

36 livraisons, price £14 12s.

GALLERIA DI TORINO (LA REALE).—
Illustrata da R. D'AZEGLIO, Memb. dell' Accad., &c. &c. Copies, Indian proofs, £18 18s.

** Bound copies in elegant half-morocco binding. India proof, £23 2s.

2 vols. folio, complete, 177 Plates of outline elevations, plans, interiors, details, &c., first impression, 140 francs, half-bound. £6 6s.

GAUTHIER, M.P., Architecte.—Les PLUS
BEAUX ÉDIFICES de la VILLE de GENES et des ses ENVIRONS. Paris, 1830-2.

Folio, 109 Plates of plans, elevations, sections, and details, £2 8s.

GRANDJEAN de MONTIGNY et A. FAMIN.
— ARCHITECTURE TOSCANE, ou palais, maisons, et autres édifices, de la Toscane. Paris, 1815.

Oblong folio, containing a profusion of picturesque views of palaces and public buildings and scenes of Venice, executed in tinted lithography, with full descriptions attached to each. Elegant in half extra morocco, interleaved, £4 14s. 6d.

KIER, G.—VENEZIA MONUMENTALE PIT-
TORESCA. Venezia.

Large folio, 61 livraisons or 3 vols., with 3 vols. of text in 4to, £18 18s.

LETAROUILLY, P.—Édifices de Rome Mo-
derne. Paris, 1825-55.

Fine Plates of the New Palace of Justice, Senato House, &c., plans, elevations, sections, doors, &c., details of the several parts, &c., £1 1s.

MICHELA, IGNAZIO.—DESCRIZIONE e
DISEGNI del PALAZZO dei MAGISTRATI SUPREMI di TORINO. Torino.

Large folio, 94 Plates, bound in extra half-morocco, gilt and interleaved, price £6 10s.

REYNAUD, L.—Trattato di Architettura, con-
tenente nozioni generali sui Principii della Construzione e sulla storia dell' Arti, con annot. per cura di Lorenzo Urbani. Venezia, 1857.

4 imperial bulky 8vo volumes, printed and published under authority, and treats of the early foundation of Venice and establishment as a kingdom, its wealth and commerce, and its once great political position, with Plates, £3 3s.

VENEZIA.—E le sue Lagune. Venezia, 1847.

VENEZIA.—Copies elegantly bound and gilt,
£4 14s. 6d. Venezia, 1847.

In 2 large folio volumes, numerously and elaborately drawn Plates, very well executed in outline, altogether a very fine work. Very elegantly half-bound in morocco, extra gilt and interleaved, £12 12s.

ACCADEMIA DI BELLI ARTI.— Opere dei
Grandi Concorsi Premiate dall' I. R. Accademia delle Belle Arti, in Milano, e pubblicate, per cura dell' Architetto, G. ALUISETTI— per la Classi di Ornano—per le Classi di Architettura, figura ed Ornato. Milano, 1825-29.

John Weale, 59, High Holborn, London, W.C.

FOREIGN WORKS, KEPT IN STOCK AS FOLLOWS:—

Atlas folio, very fine impressions, complete in 3 parts, Columbier folio, £3 13s. 6d Elegantly half-bound in extra morocco and interleaved, £5 15s. 6d.

ALBERTOLLI, G.—Alcune Decorazioni di Nobili Sale ed Altri Ornamenti. Milano, 1787, 1824, 1838.

To be had separately, £1 8s.

ALBERTOLLI, G.—Part III., very frequently required to make up sets.

2 vols., folio, 80 Plates of the most exquisite kind in colours, far superior to any existing work of the present day, £7 10s.

HOFFMAN, ET KELLERHOVEN. — Recueil de Dessins relatifs à l'Art de la Décoration chez tous les peuples et aux plus belles époques de leur civilisation, &c., destinés à servir de motifs et de matériaux aux peintres, décorateurs, peintres sur verre, et aux dessinateurs de fabriques.

Price £1 1s.

HOPE, ALEXANDER J. BERESFORD, Esq.— Abbildungen der Glasgemälde in der Salvator-Kirche zu Kilndown in der Grafschaft Kent. Copies of paintings on glass in Christ Church, Kilndown, in the county of Kent, executed in the Royal Establishment for Painting on Glass, Munich, by order of ALEXANDER J. BERESFORD HOPE, Esq., published by F. Eggert, Painter on Glass, München. The work contains one sheet with the dedication to A. J. B. HOPE, Esq., and fourteen windows; in the whole fifteen, beautifully engraved and carefully coloured.

In large folio, 80 Plates, containing a profusion of rich Italian and other ornaments. Elegant in half-morocco, gilt, and interleaved, £6 6s.

JULIENNE, E.—Industria Artistica o Raccolto di Composizioni e Decorazioni Ornamentali, come suppellettill, tappezzerie, armature, cristalli, soffitti, cornici, lampade, bronzi, ec. Venezia, 1851—1858.

Prix 50f., in folio, £3.

LE PAUTRE.—Collection des plus belles Compositions, gravées par DE CLOUX, Archte. L'Ouvrage contient cent planches. Paris.

This unique collection is in 2 Vols. 4to, had its commencement in 1812, and contains upwards of 500 rich Designs. Price £5 5s.

METIVIER, MONS., Architecte.—The original Sketches, Drawings, and Tracings, in pencil and pen and ink, of executed Works and Proposals, displaying the genius of Mons. Metivier, as an architect of high attainments, whose recent death was much regretted in Bavaria. He was a native of France, and was induced to settle in Munich by the late Duke of Leuchtenberg, under whose patronage he was much employed in the construction of private edifices for the Bavarian nobility and gentry; and for decoration and fittings of them; his interiors are still much in admiration. He built a mansion for Prince Charles, in a most simple and elegant style (in Brienner Street), which is still now considered one of the purest buildings of Munich. The above Sketches are his professional life and practice.

Twelve Parts, in small oblong 4to, 60 coloured Plates of 90 elaborately coloured and gilt ornaments, £1 1s.

ORNAMENTENBUCH.—Farbige Verzierungen für Fabrikanten, Zimmermaler und andere Baugewerke. München.

John Weale, 59, High Holborn, London, W.C.

35

In Atlas of Plates and Text, 12mo, price 25s. together,

IRON SHIP BUILDING.

WITH

PRACTICAL ILLUSTRATIONS.

BY

JOHN GRANTHAM, N.A.

DESCRIPTION OF PLATES.

This Work may be had of Messrs. LOCKWOOD & Co., No. 7, Stationers' Hall Court, and also of Mr. WEALE; either the Atlas separately for 1l. 2s. 6d., or together with the Text price as above stated.

Bradbury and Evans, Printers Whitefriars.